Bring the World to the Child

Bring the World to the Child

Technologies of Global Citizenship in American Education

Katie Day Good

The MIT Press
Cambridge, Massachusetts
London, England

This book was set in Stone Serif and Stone Sans by Jen Jackowitz. Printed and bound in the United States of America.

Library of Congress Cataloging-in-Publication Data

Names: Good, Katie Day, author.
Title: Bring the world to the child : technologies of global citizenship in American education / Katie Day Good.
Description: Cambridge, MA : MIT Press, [2020] | Includes bibliographical references and index.
Identifiers: LCCN 2019012622 | ISBN 9780262538022 (pbk. : alk. paper)
Subjects: LCSH: Education--Aims and objectives--United States--History--20th century. | Teaching--United States--Aids and devices--History--20th century. | Civics--Study and teaching--United States--History--20th century. | International education--United States--History--20th century.
Classification: LCC LA216 .G67 2020 | DDC 370.11/5--dc23
LC record available at https://lccn.loc.gov/2019012622

10 9 8 7 6 5 4 3 2 1

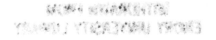

For my children

Contents

Acknowledgments

This book arose from an interest in human connections and connectivity. I couldn't have written it without the generous connections made by my mentors, colleagues, and other supportive individuals and organizations along the way.

It would have been impossible to stitch together a history of educational devices as disparate and ephemeral as these without the dedicated assistance of archivists and preservationists. I was able to find a wealth of information on nontheatrical film and screen practices thanks to the digital preservation work of Rick Prelinger and the Internet Archive. In archives and university special collections across the United States, I received helpful guidance in my search for scrapbooks, visual aids, pen pal letters, and other classroom materials. Thank you to Natalie Shilstut and Charlene Peacock of the Presbyterian Historical Society, Anne Bahde and Ben Forgard of the Special Collections and Archives Research Center at Oregon State University, Hannah Parris of the Denver Public Library, and Shirley Soenksen and Eve Measner of the Archives and Special Collections at the University of Northern Colorado. My research on the visual and peace education movements benefited greatly from the assistance of Michelle Drobik and Tamar Chute of the University Archives at Ohio State, Brandon Westerheim of the New York Public Library, Wendy Chmielewski of the Swarthmore College Peace Collection, and Vakil Smallen of the NEA Library and Archives at George Washington University.

This book began as a dissertation that was supported by a Dissertation Fellowship from the American Association of University Women and a Dissertation Proposal Development Fellowship from the Social Science Research Council. In addition to giving me time to research and write while caring for a newborn child, the American Association of University Women connected me with a network of incredible women, many of whom were current or retired educators, who invited me to speak about this project at their meetings in the Chicago area and provided incisive feedback that made it stronger. The Social Science Research Council fellowship additionally supplied me with the opportunity to develop my research in dialogue with an interdisciplinary network of faculty mentors and peer scholars. I thank William Mazzarella, Arvind Rajagopal, Andrew Brandel, Jeremy Aaron Dell, Eric Michael Hirsch, Kyle Jones, Vijayanka Nair, Bryce Peake, David Platzer, Tom Schilling, Samuel Shearer, Yana Stainova, and Niko Vicario for their close readings and invaluable suggestions on the initial stages of this book.

I benefited from a rich climate of interdisciplinary study at Northwestern University, thus allowing me to explore my interests in media, technology, representation, and history across programs and departments. I will forever be grateful for the mentorship I received from Jen Light, my doctoral adviser in the Media, Technology, and Society program. Jen sparked my interest in the history of technology, and showed me how even the most mundane of devices speak volumes about history and social relations. Lynn Spigel and Janice Radway provided additional support and encouragement in the areas of media historiography, cultural studies, feminist approaches to technology and culture, and professional development. These three women helped me understand what it means to produce scholarship on media and technology that meaningfully attends to problems of communication, representation, and power. Other faculty at Northwestern asked challenging questions, and helped me forge meaningful connections in my early work on civic technology, travel, and educational media history. Thank you to Ellen Wartella, Pablo Boczkowski, Jim Ettema, Jim Webster, Jim Schwoch, Jessica Greenberg, Jacob Smith, Mimi White, Scott Curtis, Brian Edwards, and Dilip Gaonkar for this.

I am grateful to my peer colleagues in the Media, Technology, and Society program for their friendship and feedback in and beyond graduate school. Thank you to Ignacio Siles, Rachel Plotnick, Emily Goodmann, Harsh Taneja, Robin Hoecker, Angela Xiao Wu, Heather Young, Lindsay Young, Yuli Patrick Hsieh, Brian Keegan, William Barley, Lauren Scissors, Ericka Menchen-Trevino, Aditi Raghavan, Elizabeth Lenaghan, Alan Clark, and Jacob Nelson. I have also been lucky to share in many eye-opening conversations with peers in the Rhetoric and Public Culture program, including

Kim Singletary, Caitlin Bruce, Jesse Baldwin-Philippi, Megan Bernard, Elliot Heilman, Leigh Meredith, Robert Topinka, and J. Daniel Elam. Jesse Nasta helped me make additional connections with history, and modeled how to engage critically and joyfully with the past. Finally, I am thankful to friends in the Screen Cultures program for welcoming me to their courses across campus, disciplinary conversations, and the Society for Cinema and Media Studies conference. Sincere thanks to Meenasarani Linde Murugan, Maureen Ryan, Leigh Goldstein, Nina Cartier, Zachary Campbell, Stephen Babish, Alla Gadassik, Roger Almendarez-Jiménez, Molly Schneider, Andrew Owens, Reem Hilu, Alex Thimons, Jocelyn Szczepaniak-Gillece, Dave Sagehorn, and Mabel Rosenheck.

Since transitioning into a faculty role at Miami University, my dissertation gradually morphed into a book with the help of generous institutional and collegial support. My wonderfully kind chair, Richard Campbell, provided ample time and support for me continue research and writing. The university offered summer support and a semester's worth of leave in my third year that was instrumental in terms of finishing the manuscript. The Humanities Center's Book Proposal Workshop, run by Tim Melley, helped me develop a successful proposal through conversations with colleagues from across the university. And I found all the writing motivation, inspiration, and commiseration I could ever want in my community of brilliant colleagues in the Department of Media, Journalism, and Film, including Ron Becker, Kathy German, Mack Hagood, Rosemary Pennington, Andy Rice, John Tchnernev, Hongmei Li, Kerry Hegarty, Andrew Peck, Adam Richard Rottinghaus, Carolyn Hardin, and Matt Crain. I am also thankful to Kit Hughes, Evan Elkins, Jen Malkowski, and Annie Dell'Aria for their helpful feedback and interest in this project.

Beyond my university community, I found great support for and guidance in this project from my editor, Susan Buckley, and the anonymous reviewers of this book. I am also thankful for the helpful perspectives of colleagues who've collaborated with me, and heard me present on this topic at the International Communication Association and Society for Cinema and Media Studies conferences, particularly Victoria Cain, Lisa Rabin, and Rick Popp. Thank you as well to Meryl Alper, Yu-Kei Tse, Dave Park, Jeff Pooley, Brent Malin, Rick Prelinger, Meredith Bak, Elana Levine, and Brian Gregory for their insights over the years. The project gained further definition and inspiration from the dedicated work of Josh Shepperd, who invited me to take part in the Radio Preservation Task Force at the Library of Congress in 2015 and helped me find a place for my work in the history of broadcasting. I am also grateful for my conversations with supportive academic mentors from my undergraduate years, including Kathleen Adams and John Terrell, and the colleagues I have connected with online, including Michael Socolow, Andrew Schrock, Talya Zemach-Bersin, and Eileen Clancy.

Most of all, I am grateful to my family for supporting me with love, time, resources, and childcare over the years. My parents, Tom and Marcia, unknowingly instilled in me a lifelong fascination with technology, pen pals, and multimedia when they brought home a Gateway computer in 1995. When virtual contact with the wider world was no longer enough for me, they sacrificed a great deal so that I could travel abroad to Ecuador as a high school exchange student—an experience that would profoundly change my worldview and open innumerable opportunities to me. While there, my Ecuadorian host families, Marcelo, Sagrario, Andrea, and María Leonor Noboa, and Vicente, Laura, Yeseña and Omar Díaz, along with my dear friend from Germany, Janna Rakowski, broadened my cultural and linguistic horizons in ways that have shaped my life and learning ever since. The experience helped me to later secure two scholarships funded by the State Department to study abroad again during and after college—first as a National Security Education Program's Boren Scholar to India and later as a Fulbright-mtvU Fellow to Mexico. These travels were deeply instructive and enlightening for me, and prompted me to examine how my own formation into a world-minded US citizen was shaped by problematic histories of race, representation, education, and statecraft. These questions led me to take up the research for this book.

I am above all grateful to my husband, Alejandro, who reminded me of this project's worth whenever I had doubts about it. I could not have finished it without his emotional support and loving parenting of our two sweet boys, Nico and Emilio. This book is dedicated to them.

1 A Wonderland of Devices

Our days of "splendid isolation" are past. The American child of today and tomorrow is a world citizen.

–Anonymous, *Visual Education* magazine, 1921

In 1924, James N. Emery, an elementary school principal in Pawtucket, Rhode Island, penned an essay in praise of visual aids for *Educational Screen*, a new journal dedicated to the use of visual media and technology in education. "Modern science has opened a wonderland of devices to make education real and vital to the boys and girls in our school rooms," he began. "If the boy in tippet and mittens, trudging over the muddy road to the little red school house of fifty years ago, had been set down in one of our great modern buildings, he would have refused to believe his eyes." Emery went on to list an array of modern "visual helps" that teachers were beginning to incorporate into their lessons, emphasizing that they included not just motion pictures, the most talked-about technology of the day, but also lower-profile "visual devices of all sorts; among them the lantern slide, the photograph, the half-tone and rotogravure picture, the stereoscopic view, the chart, map, globe, sand-table, outline map, the mineral and product collection, the model, and even the pageant."

To demonstrate the modernity and pedagogical power of these diverse media and materials, Emery explained how some schools in Rhode Island were using them to bring the world into the classroom. Students viewed lantern slide images of Africa, Siam, and the South Pole, collaboratively assembled bulletin boards and scrapbooks with pictures clipped from *National Geographic* magazine, and created exhibits out of products from "the Orient." For Emery, these devices were important additions to the classroom because they brought worldly experience to the untraveled. "From very lack of travel and experience," he wrote, "the helps that visual aids can give are most sorely needed." By democratizing a world of knowledge, he concluded, these sensory aids and activities would help students "to be better prepared for the complex duties of modern life, to be better citizens, better Americans."[1]

Emery's account testifies not only to the proliferation of educational technologies in schools in the early twentieth century but also to the emergence of a popular American belief that equipping schools with such devices could prepare young people for citizenship in a more globally interconnected and mediated future. Long before educators, industry leaders, and policy makers would tout the importance of "global," "wired," and "multimodal" learning in the digital age, their predecessors in the machine age were already describing a suite of new communication technologies and techniques as tools to "bring the world to the child" and expand the cultural horizons of youth. Reformers praised the world-channeling capabilities of not only new mechanical media—such as film, radio, and lantern slides—but also humbler, more grassroots media of communication created by teachers and children, including pageants, object collections, and pen pal correspondences. By teaching with these varied "devices," they suggested, educators could transform the public school from the passive, provincial, and textbook-bound schoolhouse of the nineteenth century into a modern, multisensory, and cosmopolitan center of learning. New sensory aids and activities could efficiently familiarize young Americans with distant people, places, and processes in ways that would leave them better prepared to participate in a more pluralistic democracy, mass-mediated society, and industrial and international economy.[2]

This book explores how a variety of emerging and existing media technologies entered US schools in the first half of the twentieth century under the banner of promoting cosmopolitan and civic-minded learning. What today we call "connected learning" and "global classrooms" are generally regarded as new ideals in education, driven by the rise of digital technology, the internet, and the global information economy in the last few decades.[3] Efforts to historicize these ideas usually look to the 1960s, when the Canadian communication scholar Marshall McLuhan famously proclaimed that new "electric media," such as television, satellites, and computers,

Figure 1.1
A classroom of "geographic enthusiasts" in Saint Cloud, Minnesota, learns geography with the help of a variety of media and technologies, including models, maps, globes, *National Geographic* magazines, photographs, and a lantern slide projector. From Jessie L. Burrall, "Sight-Seeing in School: Taking Twenty Million Children on a Picture Tour of the World," *National Geographic* 35 (1919): 500.

were transforming the world into a "global village" and "classroom without walls."[4] McLuhan's widely reproduced claims helped fuel decades of American investment in educational technology as part of a broader Cold War strategy to compete with the Soviet Union and dominate in the global information age.[5] In the twenty-first century, educational technologies are still widely regarded as tools that can advance the "universal" ideals of democratic communication and international understanding, while also supporting the economic and geopolitical interests of the United States. But where did these assumptions come from? When did mediated communication in the classroom come to be associated with advancing both world peace and American power?

I argue that American visions of a media-enriched classroom and world-minded citizenry emerged and converged with each other in the first half of the twentieth century, amid rising social concerns about intercultural contact, mass media, and global war. This period saw both a surge and sudden curtailment of immigration to the United States, the migration of racial and ethnic minorities to cities, the rapid expansion of the public education system, two devastating world wars, wonder-inspiring advancements in international communication and transport, and an influx of new media technologies into popular culture, homes, and schools. For the white progressives and reformers who ran the public education system, these changes prompted a mix of concern and optimism about the role of the neighborhood school in a global age. Observing their "shrinking world," their discourses revealed anxiety about new media's influence over children, an idealistic interest in harnessing new media to teach national unity and international understanding, and a broadly shared conviction that the United States was an exceptional nation that could model modernity and pluralistic democracy for the rest of the world.

By bringing a mix of communication tools and techniques into the classroom, educational reformers and technology producers became powerful arbiters of the circulation of global and intercultural imagery as well as ideology in American life. In their view, instructional media was valuable not solely for enhancing teaching and learning but also for assisting children in looking outward to a fast-changing world as well as inward on their own national greatness and social progress. The vision of mediated cosmopolitanism that they embraced, in short, showed early signs of the late twentieth-century neoliberal mind-set that the geographer Katharyne Mitchell has called "strategic cosmopolitanism"—or "cosmopolitan learning in the service of the national interest."[6] By endowing new communication devices and practices with the capacity to "bring the world to the child," reformers and technology promoters believed they could mold diverse and provincial youths into world-minded, but also thoroughly American, citizens.

This book aims to trace a critical history of not only educational technology but also the K–12 school as a rich domain of global media reception, production, and exchange in American everyday life. Emerging in the nineteenth century, the public school system was arguably the "first mass medium" in the United States—an institution designed to efficiently disseminate information, images, and ideology to entire generations of young people.[7] The early twentieth century saw its rapid expansion alongside a burgeoning mass communications sector, with both serving, alongside YMCAs, factories, museums, department stores, parks, and other emerging spaces of community interaction and mass information, as "crucibles of modern subjectivity" for a growing, urbanizing, and increasingly diverse population.[8] Yet the early uses of audio and visual media in classrooms have received less attention from historians and critical/cultural scholars than has film, broadcasting, the press, or popular visual culture. School was the place where most American children first began to form a formalized idea of the international world and their role within it, yet many of the pedagogical devices used to teach these ideas, such as stereoscopes, pageants, and pen pal letters, have yet to be incorporated into the history and scholarship of global media. Debates about educational media and technology, too, were vibrant discursive spaces in which ideas about the role of multimediated, global, and participatory communication in a modern democracy were forged.

In exploring this history, I aim to make three contributions that bridge the fields of educational history, media and communication studies, and critical and cultural studies of technology. First, I argue for accounting for the diversity of technologies that have historically been used and made in schools, particularly to teach children about cultural difference and the international world. Second, I call for recognizing schools as an active, influential, and problematic circuit of global and intercultural media and communication in American everyday life. Finally, I examine the social and political ideologies that have historically underpinned the educational uses of media to teach good citizenship in a globalizing age.

A Diversity of Educational Technologies

First, understanding how global information and images circulated in twentieth-century schools requires an expansive view of educational technology—one that is inclusive of both the high-profile devices that were peddled by technology companies and chronicled in the press, and the lower-profile, more mundane media and materials that adorned classroom walls and filled school cabinets and closets. While virtually any informational device can be conceived of as an educational technology, this book defines educational technology and instructional media as the visual, audio, and sensory devices that educators experimented with as an alternative, or supplement, to textbooks and verbalism (the written and spoken word). These devices flooded into schools in the twentieth century, ushered in by progressive educational reforms that emphasized the development of the "whole child" through the creation of learning environments and lessons that engaged their individual interests and senses as well as the visual education movement (discussed in chapters 2 and 3), which strove to enhance instruction with visual, audio, and other sensory aids.

Historical and contemporary narratives about educational technology, like those describing technology in general, often follow what David

Edgerton calls "tidy timelines of progress," focusing on the introduction, successes, and failures of "big, spectacular, masculine high technologies," such as film, radio, teaching machines, television, computers, and digital tablets, in schools.[9] This progress narrative mirrors the promotional rhetoric of the educational technology industry, which has long positioned its products as singular, stand-alone, and "revolutionary" aids that can serve as instructional panaceas. Thomas Edison, an early producer and promoter of educational films, was one of the first and most famous progenitors of this techno-idealistic rhetoric in American society. In the late 1910s and early 1920s, he declared, to much controversy, that the motion picture was "destined to revolutionize our educational system and that in a few years it will supplant largely, if not entirely, the use of textbooks."[10] A century later, the nation's most powerful technological tycoons, from Bill Gates to Mark Zuckerberg, along with a growing, multibillion-dollar "EdTech" industry, continue to make headlines with claims about the potential of high-tech innovations to "disrupt" and transform education.[11] After decades of these discourses, the term "educational technology" is more likely to conjure images of a sleek, cutting-edge, electronic consumer device—such as a SMART Board or iPad—than a child's drawing, scrapbook, or drawer full of rocks.

Yet in most schools, the integration of technology has been far from tidy, and no single device has yet managed to significantly upend established educational practices and tools.[12] From the stereoscopes and lantern slide projectors of the early 1900s, to the computers and virtual reality viewers of the early 2000s, commercial technologies have always been distributed and used unevenly in schools. Teachers have long complained of technologies being foisted on them by higher-ups, with little time and money allocated toward training them on how to use them effectively.[13] Moreover, each new gadget introduced into the classroom has been likely to share space and status with a vast array of "old" media and humble informational devices that have remained largely unnoticed by researchers and reporters, and in some ways, unchanged for over a century. These low-tech, sensory standbys, many of them prized by educational professionals for the relative ease with which they can be collected, created, or modified to suit their instructional needs, include object collections, bulletin boards, erasable boards, scrapbooks, artwork, posters, charts, maps, exhibits, and, as Emery noted at the beginning of this chapter, even performances by students, such as pageants.

Recognizing the historic plurality and participatory nature of instructional media in the classroom underscores the enduring role of women and children—figures traditionally left out of histories of technology—in shaping the transmission of mediated information and imagery in school and society.[14] It also responds to calls from scholars of media studies, science and technology studies, and critical information studies to take a more material and "intermedial" approach to studying the role of technology in society. This approach acknowledges that media technologies operate neither in isolation from each other nor according to preordained use values, but rather in dynamic relation with other devices and objects, in daily practices of use and disuse, and embedded within complex social, political, and material contexts. As Estrid Sørensen has argued, learning with media in the classroom is not only a social but social-material process too, as novel and emerging school technologies (e.g., weblogs, conference systems, and online virtual worlds) are usually used in dynamic combination with established ones (e.g., blackboards, bells, paper, students' bodies, and chairs), often contributing to learning in ways that diverge from the expectations of their creators, implementers, and users.[15]

For many educators in the first half of the twentieth century, deploying multiple types of sensory aids and activities in the classroom, rather than relying on a single commercial device, was not merely a matter of convenience or personal preference but also an expression of progressive ideals of liberal, multisensory, and student-centered learning. The educators who advocated for teaching cosmopolitanism, in particular, took a distinctively cosmopolitan approach to educational technology, frequently listing litanies of devices that could be mixed and marshaled together to support world-minded learning. For example, Edgar Dale and Rachel Davis DuBois, two influential interwar educational researchers (discussed in chapters 3 and 4), urged schoolteachers to make creative use of multiple forms of media— including films, radio programs, newspapers, and school assemblies—to teach American youths about foreign cultures, and train them to detect and dispel stereotypes. Both saw these engagements with diverse forms of information as necessary for "counteracting" the totalizing influence of an expanding mass media sector as well as "inoculating" young people with critical, democratic, and world-minded attitudes to resist the pull of fascist propaganda. The development of this worldly, participatory, and multimedia sensibility in pre–World War II pedagogical writings and practice was foundational to the later twentieth-century development of multimedia in educational spaces, and the American enthusiasm for "wired" classrooms along with "global" and "connected" learning.[16]

The School as Circuit of Everyday Global Mediation, Communication, and Contact

The second contribution of this book is to turn a lens on the school, particularly the public school, as a historically active, vibrant, and fraught domain of global and intercultural media reception, production, and circulation in American life. The early twentieth century is rightly remembered as a time of intense nativism, jingoism, and racism in the United States. Between 1871 and 1924, over twenty-four million immigrants, including large numbers of Italians, Jews, and Poles, arrived in the United States—a number larger than that which had arrived in the previous two and a half centuries combined.[17] By the 1910s, African Americans began migrating northward and westward from the Jim Crow South, fleeing racial terror and seeking economic opportunities in industrial cities. The increased contact among racial and ethnic groups, described by John Dewey in 1902 as a "sudden massing of heterogeneous elements," prompted the white, native-born Protestants who controlled the reins of education and government to look for new ways to preserve the Anglo-American character of the country.[18] Educators and employers developed programs to "Americanize" immigrants and educate away their old-world ethnic identities. Congress passed the Immigration Act of 1924, which drastically reduced immigration from outside northern Europe and excluded Asians altogether. The Ku Klux Klan resurged in cities and towns across the United States, promoting an ideology of white supremacy, discrimination, and violence against African Americans, immigrants, Catholics, and Jews. And both of the world wars intensified pressure on schools to teach nationalism, patriotism, and "unity."[19]

But alongside these forces that reinforced white hegemony and cultural conformity, the first decades of the twentieth century also saw, paradoxically, a budding mainstream curiosity about cultural diversity and the international world. In urban schools and settlement houses, many educators and social workers, some of them sympathetic to the plight of immigrants or recent immigrants themselves, adopted a rhetoric of "patriotic

pluralism," advocating not for the draconian Americanization of immigrants but instead a celebration of their traditions as essential to the US
story.[20] Additionally, as Kristin Hoganson argues, the expansion of American empire abroad in the form of trade, tourism, and newly occupied territories made global products and experiences available as well as desirable
to a growing middle class. Literary and popular culture too chronicled the
"tension between small-town insularity and yearnings for a connection to
a wider world."[21] With rapid advances in air travel, wireless technology,
and global communications and transport, observers described the 1910s,
1920s, and 1930s as an "age of super-contact" and the birth of a "com-
pact, cohesive, world society."[22] "Steam and electricity have made the world
smaller," declared Julia Brown, an English teacher, in 1934. In this interconnected age haunted by old prejudices, she asserted, teachers had a special
duty to "develop an international viewpoint in children" to ensure a more
peaceful future:

> New York is closer to Europe than it was to its own suburbs a few years ago. For
> practical purposes the world really is getting smaller, and, in this small world, the
> lives of men and women in different countries touch one another in a thousand
> ways for every contact they had a hundred years ago. This makes it imperative
> for prejudices to be thrown away and understanding and sympathy made to take
> their place. The future peace of the world depends largely upon the growing num-
> ber of world-minded citizens and upon a peace-minded public in every country
> of the world.[23]

By the eve of World War II, it was common for American children and
adolescents to encounter vivid, and often deeply racist and imperialist,
representations of cultural difference not only in a booming mass culture
of movies, radio, and print but also in new school-sanctioned media and
activities intended to mold them into tolerant, unprejudiced, and "world-
minded citizens."[24] Since many teachers lacked intercultural or international experience themselves, they turned to stereoscopes, lantern slides,
films, radio programs, bulletin boards, *National Geographic* materials, international pageants, exhibits, and pen pal programs to help them in this task.
Many enlisted their own students, the descendants of immigrants, to mine
their family's closets and heirloom cupboards for cultural garb and "curios"
to bring to school.[25]

While the public school system remained deeply segregated by race,
with Black,[26] Asian, Latino, and Native American youths barred from most

mainstream schools and sent to severely underfunded ones, well-resourced white schools increasingly, and ironically, made room for international images, objects, and media to shape their students into future-ready cosmopolitans.[27] Embracing the idea of bringing a generalized, friendly world to children with the latest technologies and educational principles, educators and audiovisual (AV) advocates could perform their commitment to preparing children for citizenship in a diverse, interconnected society without attending to the challenges of racial segregation and inequality in their own communities. For Pauline Powers, a geography teacher in Massachusetts writing in 1928, the availability of new audio and visual media meant that students could now cultivate their interests in foreign and faraway lands in ways previously not possible in the olden days of rote, textbook-based instruction:

> The desire to learn about other peoples and other lands as well as to know our own more intimately grows in every human heart, but in other days it did not flower. One did not sail thru South Eastern Europe to the strains of The Beautiful Blue Danube on the schoolroom Victrola; one was too busy memorizing the location of the Danube's source . . . and anyhow there was no Victrola! No Victrola, no radio, no motion picture, few newspaper or magazine articles of geographic interest in those days! But how many aids there are at the disposal of the geography teacher of today![28]

Acknowledging the presence and variability of cosmopolitan media texts and practices in classrooms nuances prevailing narratives about the history of global media and communication in the twentieth century. Histories of global media have primarily focused on mediators who were male and professional, and large-scale circuits of colonial administration, geopolitics, commerce, and popular culture.[29] They have also centered on the "outgoing tide" of mass-produced, American-made mass media products, news, propaganda, and cultural influence that the United States exported to other countries as a form of soft power.[30] Less attention has been paid to the production and circulation of internationalist media within the United States along with the historical reception of these media within the domestic sphere, particularly among children and in the citizenship-building institution of the schools.[31] The global and internationalist media that circulated in American schools was a complex mix of texts produced by influential institutions—such as Hollywood and nontheatrical films, state- and federal-supported radio programs, and images from the National

Figure 1.2
Advertisements for educational technologies frequently linked products to discourses of virtual travel, exploration, interest learning, and preparing children for the future. *Educational Screen* 28, no. 6 (1939): 217.

Geographic Society—as well as more grassroots, localized, and amateur forms created by individual teachers and learners. A notable example of the latter, discussed in chapter 5, was the tens of thousands of pen pal letters, scrapbooks, dolls, and care packages exchanged internationally among children after World War I under the auspices of organizations such as the National Peabody Bureau of International Correspondence, Committee on World Friendship among Children, and Junior Red Cross.

Recovering these worldly media "mixes" in schools, praised by educators, industry leaders, and policy makers alike as symbols of a modern, democratic, and child-centered education, can shed light on today's enthusiasm for harnessing digital technologies to teach "new literacies" and "intercultural competencies" to children. It can also contribute to what David Morley has called a "materialist, non-media-centric" study of media globalization.[32] Rejecting what he summarizes as the "abstracted 'One-Size-Fits-All' analyses of globalisation-thru-technology," or tendency to study global communication through the lens of high-speed, high-tech, and deterritorialized flows of information, Morley argues instead that "global cultural forms still have to be made sense of within the context of local forms of life." Calling for a more "grounded" approach to globalization that entails a return to the "classical" definition of communication—one that accounts for not only the virtual transmission of information but also the physical transport of people and things—Morley asks that we "incorporate the study of the movement of people, and commodities, and technologies alongside that of information—and contemporaneously, place the study of the emerging virtual realms alongside that of the material world."[33]

This study does not attempt to provide an exhaustive account of global media technologies in American classrooms. Drawing from archival teaching journals, teacher training textbooks, newspaper reports, and educational and media archives, it can only scratch the surface, and attempt to give a sense of the variety of media talked about, advertised, and used by educational researchers, administrators, and teachers in schools. While evidence of world-minded discourses and AV initiatives can be found in schools across the United States, they were most common in urban areas and their well-heeled suburbs in the Northeast and Midwest, and the uniquely progressive and multicultural school systems of New York City and Los Angeles. The definition of "world-mindedness," furthermore, was capacious and vague, leading it to be attached to educational programs and texts that variously sought to acknowledge and combat prejudice, and in other instances, reinforce racist and imperialist orderings of the world. It is also important to note that while this book highlights a persistent and intensifying hum of outward-looking inclinations in educational and industry rhetoric, it is difficult to know the extent to which such ideas were actually implemented in day-to-day instruction, with the exception of the specific examples that I highlight throughout the book. Scholars of education have long noted tensions between hoped-for reforms and the realities of everyday instruction, in which teachers have typically taken a more conservative and traditional approach to teaching than that recommended by idealistic pedagogues.[34]

Additionally, this book does not address, due to the limitations of time and space, the important twentieth-century technologies of filmstrips, which are long overdue for historical analysis, and television, which has been examined elsewhere, and offers only a cursory look at the vast fields of educational film and radio.[35] Nor does it claim that world citizenship was the only or even most important ideal in facilitating the uptake of technology in schools, as appeals to efficiency, economizing instruction, and making lessons more interesting and memorable were made just as often, if not more, in the educational literature of this period. This study, however, does endeavor to show how today's internationalist and techno-utopian rhetorics and logics in education have deeply historical roots, and extend from a long legacy of efforts to equip schools as well as shape particular kinds of citizens and worldviews with emerging media. By the end of this book, I hope the reader will have a sense of how vast and varied mixtures of high and low, professional and amateur, altruistic and problematic technologies have long shared in the work of bringing the world into the neighborhood school.

Ideologies of School Media

Finally, this account examines the ideological ambiguities and assumptions that undergirded the promotion, proliferation, and use of global media in schools. A growing body of research has shown that mass-produced educational media, including textbooks, motion pictures, television, and computer games, have long reinforced dominant social norms and values in American schools and society, including assimilation, Christianity, free market capitalism and industrialism, individualism, militarism, racism, imperialism, and anti-Communism.[36] But more attention is needed to understand how such ideological messages and representations cut across mass-produced and amateur media forms as well as how these instructional media were mobilized in the service of teaching normative definitions of citizenship to children through representations of cultural, racial, ethnic, and national difference. As Victoria Cain has observed in her study of lantern slides, stereoscopes, and other visual aids in early twentieth-century geography instruction, despite the heterogeneity of media production and notorious decentralization of the public school system, "a remarkably consistent educational iconography of the globe nevertheless emerged between the 1900s and the 1920s," offering a "tightly constrained global taxonomy of discrete objects, places and types, one whose scope was framed by curricular emphasis upon physical land-forms, commercial products, and racial-cultural difference." These and other sensory media, whether made by educational companies or schoolteachers clipping pictures from magazines, encouraged children to view the world as "mappable, nameable, and knowable in its entirety," with each nation reducible to a few notable cultural traits, industries, and landmarks (associat-ing the Dutch, for example, with wooden shoes and windmills).[37] Some edu-cational media, such as the National Broadcasting Company's (NBC) natural history radio program, *The World Is Yours* (discussed in chapter 2), rendered the world as not only accessible and knowable but also virtually traversable and *possessable* by ordinary citizens through the space- and time-bending superpowers of new media technology.

For E. Winnifred Crawford, a New Jersey educator and early advocate of visual education, transforming the classroom into a space "splendidly equipped with illustrative material" was vital to ensuring that children understood their roles as citizens and interdependence with other nations. "A great company of cosmopolitan people live in America. We have the world in America," she said, speaking before the Visual Education Club of Newark's public schools in 1921. "We must teach them the world in order to make them Americans. This we can better accomplish by using visual aids." Crawford, who went on to develop one of the first teacher educa-tion courses in visual education, claimed that by teaching with a variety of visual devices, including "lantern slide, stereograph, picture, curio, and model . . . we can so picture the contributions that each race and each country has made to our civilization that the children will have a greater respect for all peoples and that 'dago,' 'wap,' and 'chinney' will spring less often from their lips." In addition to teaching students to appreciate "the brotherhood of man," she argued, visual aids could introduce them to the unique wonders of their own nation, such as the Rocky Mountains and Niagara Falls, and thus instill in them a national pride and willingness to defend America's democratic freedoms:

[Visualized] geography can contribute as much, if not more, towards the Americanization of our heterogeneous population than any other subject in our curriculum. We can so teach that the children will love their country with its beauties and wonders so much that they will work to make a strong democracy develop within her. Then they will desire only those things that will keep their loved land safe. This comes to us as a challenge to use visual aids in teaching geography that those under us will never become partners to anything that will bring disaster to their land or its government.[38]

Through claims like these, boosters of audio and visual education contributed to ambivalent understandings of educational technology as diverse, democratic, and socially uplifting as well as politically and economically useful. They contributed to an emerging, uniquely twentieth-century vision of the public school as not only a medium of mass instruction and citizenship formation in its own right but also a richly mediated space where children could develop tolerant yet patriotic subjectivities through cosmopolitan uses of multiple types of media.

Global educational media were thus produced and promoted as a means to both expand and carefully control the representation of the world in the classroom. For some educators, bringing audio and visual aids into the classroom offered a way to "counteract" the harmful social messages, propaganda, and prejudice-forming stereotypes that circulated in mass culture, while still hewing closely to dominant definitions of race and respectable citizenship. When they noted that new technologies could, as one visual educator claimed, "reveal the races of the world to one another," and according to another, "bring about a mutual understanding that will further the brotherhood of man and promote Universal Peace," most were interested in teaching only a *generalized* global awareness that did not disturb the existing racial hierarchy or narratives about the supremacy of white, middle-class, US citizenship.[39] As the following chapters will highlight, while many teachers enthusiastically embraced new instructional media to help children become more "world-minded," few used these devices to illuminate the entrenched problems of structural racism or segregation within their own schools or society.

By wielding a rhetoric of mediated cosmopolitanism and techno-idealism, early twentieth-century educators, media producers, and government officials breathed new life into ambiguous Western discursive traditions of characterizing media devices as neutral, scientific, and "universal" instruments that could reliably compress, contain, and channel the sights and sounds of a diverse world for the edification of the student or spectator. The notion of media serving as a realistic lens or "window" into faraway places is a legacy of European tools of colonial exploration and exhibition, such as telescopes, microscopes, magic lanterns, museums, and cabinets of curiosities.[40] As the art historian Barbara Stafford and curator Frances Terpak write of these "capturing devices," which were developed by and for the benefit of white, affluent Europeans, "They gathered together divergent messages received from the cosmos and projected them in brighter, sharper form . . . and concentrated, conjoined, and metamorphosed the universe's unruly singularities."[41]

Such developments in the world of visual technology dovetailed with an outward-looking interest in Western education, where efforts to neatly organize and display the "world in a box" have manifested in a range of educational materials from at least the seventeenth century to the present. These include some of the earliest illustrated textbooks and geography readers, such as John Amos Comenius's *Orbis Sensualium Pictus* (1658); the "imperial displays" of the eighteenth and nineteenth centuries, including museums, botanical gardens, and zoos; nineteenth-century object lessons and maps; and children's media in the early twentieth century that presented the idea of internationalism in miniature through stories about model children in different countries, as seen in Lucy Fitch Perkins's *Twins of the World* book series along with the educational book and film series *Children of All Lands*, produced by Pathé Pictures in 1928–1929. The legacy of condensing a complex, changing world into child-friendly forms and selling media products as tools of fictive travel and education has endured in the twenty-first century. One popular example is Little Passports, a mail-order educational subscription service that promises parents it will "inspire your child to learn about the world" through monthly deliveries of international-themed toys, games, maps, and a "pen pal letter" from fictional children in faraway lands as well as worldly stamps and stickers to decorate a handsome passport and suitcase.[42]

The notion that media can efficiently "capture" the world, compress its most useful or educational contents, and tame its "unruliness" for the enlightenment and enjoyment of the user has also persisted throughout the history of marketing of modern media technologies to the general public, from early visual devices to radio, television, computers, and smartphones. "Bringing the world" to the spectator has proven to be a particularly useful discourse for promoters who wish to demonstrate the instructive, democratic, or socially uplifting potentials of technological novelties. Magic lantern exhibitions in the late nineteenth century, for example, promised to take upwardly mobile Americans "Around the World in 80 Minutes" through a combination of projected travel photography and lectures by esteemed explorers. Over half a century later, as Lynn Spigel has shown, postwar promotions for the new technology of television celebrated the capacity of the medium to "bring 'another world' into the home," figuring TV as "the ultimate expression of progress in utopian statements concerning 'man's' ability to conquer and to domesticate space." Technology's promise to democratize global exploration and virtual travel continues to hold sway in digital culture, where metaphors of web "navigators" and an "information superhighway" have contributed to what Lisa Parks calls a "fantasy of digital nomadism" and "global presence" in promotions of the internet.[43] On the one hand, the recurrence of such utopian claims across time and technologies testifies to the endurance of Enlightenment era ideals of harnessing technological innovations to more fully connect, represent, and understand the world. But it also hints at the darker side of media utopias, particularly the Western imperialist impulse to explore, condense, and represent the world's cultural, geographic, and biological diversity for purposes of classification, consumption, erasure, and control.

Technologies of Global Citizenship

The title of this book, *Bring the World the Child*, was the slogan of the St. Louis Educational Museum (SLEM), one of the first initiatives to bring a variety of media technologies into public schools at the turn of the twentieth century. It is worth taking a moment to describe this unique educational service, which helped launch the visual education movement, because it encapsulates several of the features common to the pedagogical devices I will discuss in this book. Specifically, the SLEM is an example of what I am calling "technologies of global citizenship" in American education. Technologies of global citizenship refers to the constellation of media, materials, and activities that educators adopted to promote worldly yet hegemonic learning.

Established in 1904 by the Board of Education in Saint Louis, Missouri, the SLEM was a mobile school museum service that began as a collection of approximately five thousand objects, including cultural artifacts, natural specimens, films, lantern slides, boxed exhibits, costumes, and samples of industrial products and foods. In what might be considered a predecessor to the modern AV cart or mobile media library, this traveling museum, nicknamed the "museum on wheels," began with a horse-drawn wagon, later exchanged for a fleet of trucks, that delivered crates of educational materials on loan to public school teachers throughout the city. The museum was created with artifacts left over from the Louisiana Purchase Exposition, the city's elaborate world's fair of 1904, and later enhanced with items donated from the Field Museum of Natural History in Chicago, the Smithsonian Museum in Washington, DC, and public museums in Philadelphia and Milwaukee.[44] The SLEM's varied and worldly collections made it the embodiment of emerging progressive ideals of world-minded, sensory, and active learning. Promoters of the SLEM boasted that bringing a diversity of materials into schools could effectively liberate children from the failed pedagogy of rote verbalism and "bring the world" to those who lacked the means to travel. As Carl Rathmann, the superintendent of the Saint Louis schools and director of the museum, wrote,

> We are being freed from the idea that the textbook is the only means of teaching, the only source of information. . . . We are beginning to realize that to make the child acquainted with the world in which he lives we must bring him into personal contact with the world. We must . . . "take him into the world or bring the world to him." We can do this through the school museum with a stock of well-selected and carefully arranged material.[45]

Like the world's fair on which it was modeled, the SLEM showcased the cultural, natural, and industrial diversity of the international world through an imperialist lens of American expansion, progress, and power. Children were encouraged to "explore and discover" the collection with their senses, and in the process, learn about their own advanced "place among the nations." They could trace the global journey of raw industrial resources like rubber, for example, by handling samples of manufactured tires and toys, studying a wall chart depicting a rubber tree, and viewing three-dimensional stereograph images of the "geography, the people, homes, costumes and customs of the far-away lands from which the world gets its rubber supply." In another unit, children could put on Native American costumes and make teepees, arrows, foods, and pottery, discovering

Figure 1.3
"The schoolroom turned into a workshop portraying Indian life." In Carl G. Rath-
mann, "Visual Education and the St. Louis School Museum," *Bulletin of the Bureau of
Education, US Department of the Interior* 39 (1924): 2.

through embodied experience the contrast between their own modern existence and the ancient ways of "Indian life."[46]

By 1914, the museum captured the attention of US commissioner of
education Philander P. Claxton, who praised its ability to bring the benefits of a modern education to the underprivileged. "For any accurate ideas
of the things of the world at large, the child must be taken on extensive
journeys or the things of distant places must be brought into the school.
For most children the first is clearly impossible."[47] The museum was so
popular in Saint Louis that by the 1920s, it expanded to occupy an entire
building—a former school—and reportedly contained over 10,000 collections of objects, 20,000 lantern slides, 25,000 stereographs, 1,200 colored
wall charts, 800 phonograph records, 250 reels of film, thousands of photographs, and countless books. To educational observers, this massive collection of media and materials was the ultimate educational technology:

an encyclopedic, recombinant, multisensory, mobile, and public resource designed for active use in the service of world-minded learning.

The SLEM enshrined a budding vision of mediated pedagogy that was distinctively civic, global, multisensory, participatory, and ethnocentric. It was *civic* insofar as it was understood as a tool for instructing, not entertaining, the public, and preparing children for responsible citizenship and productive labor in a changing, increasingly global and interdependent society. It was *multisensory* insofar as it contained multiple types of visual, audio, textual, olfactory, and tactile media designed to engage children through all the senses. It was *participatory* insofar as its pedagogical value depended on students actively handling, exploring, assembling, recombining, and interacting with its contents. And it was *ethnocentric* insofar as it positioned the United States and white, Anglo-American identity at the center of mediated lessons on the international world, and exhibited cultural differences in ways that reinforced, in ways subtle and overt, colonial, Euro-American ideologies of racial and national superiority and progress. Though developed over a century ago, the SLEM was a cutting-edge educational technology that anticipated many of the contradictions, potentials, and problems that have persisted in the internet as well as other digital and electronic technologies in education today.

The SLEM was emblematic of an emerging crop of tools and techniques in early twentieth-century American education that can be conceptualized as "technologies of global citizenship." I use this term to draw a circle around the heterogeneous and dynamic assemblages of media, materials, and embodied practices—or what teachers simply called "devices"—that were used not simply to teach but also to help mold youths into world-minded and patriotic US citizens. In the first decades of the century, as the United States shifted from a stance of isolationism to internationalism and took on a more active role in world affairs, educators called for teaching a new kind of citizenship that included a sense of connection and responsibility to not only the local community and nation but the entire world. This new, broadened ideal of "world citizenship," variously referred to as "world friendship," "world-mindedness," and "international understanding," was, according to William Carr, a researcher for the National Education Association (NEA), a "type of citizenship wider than the boundaries of nations." Education for world citizenship, he explained in a 1928 book of the same title, was "education for sympathetic, peaceful co-operation based on democracy."[48]

For many, this educational imperative was a necessary response to the proliferation of new technologies of communication and transport, which made the world feel "smaller" while increasing opportunities for contact, commerce, and conflict between nations. "The enlarging of fellowship of human life upon this planet, which has moved out through ever-widening circles of communication and contact, has now become explicitly and overwhelmingly international, and it never can be crowded back again," wrote Ercel McAteer, who directed visual education services in the Los Angeles City schools, in a 1928 essay on the importance of educational motion pictures in fostering "international co-operation." Stella Center, the president of the National Council of Teachers of English, remarked in a speech in 1933 that the growth of new technology, along with the devastation of the First World War, had made it clear that American schools should be working to "develop a feeling of world solidarity and create a better international understanding" in children:

> The parochial, the insular, the sectional, the narrowly nationalistic attitude, is an anachronism in a world growing rapidly smaller by means of the airplane, the radio, the telephone, and television. Every Main Street in America was shocked and dazed by the murder of an Austrian archduke in a remote city on the border of Serbia, in 1914. The World War enlightened the thinking American and hastened the passing of parochial-mindedness. The political effective of the World War was to enlarge the horizon of the common citizen.

A similar assessment came from William Kilpatrick, the prominent progressive pedagogue and student of Dewey's, in an address at Rutgers University in 1926:

> As men and their affairs become more and more interrelated, each individual with other individuals, one groups with other groups, one nation with other nations, there is a corresponding demand for an outlook adequate to take care of this far-flung and growing connectedness. Nothing less than world-mindedness will suffice—the ability to see social problems on the scale on which they exist.[49]

But in a time of intense nativism, xenophobia, and anti-Communism, advocating for teaching world citizenship in schools was risky. Patriotic and hereditary societies, such as the Daughters of the American Revolution (founded in 1890) and American Legion (founded by veterans of the First World War in 1919), functioned as public school watchdogs, closely monitoring textbooks, curricula, and teachers for any mention of ideas that could be interpreted as unpatriotic or "un-American." In this climate, many educators who favored teaching world citizenship were careful to do so in an apolitical and "indirect" way. "We pass lightly over the details of wars," explained one junior high teacher in the peace education handbook *Educating for Peace* in 1930. "We try to show that the wars themselves were unnecessary. We stress the Kellogg Peace Pact and other movements for world peace. And I personally do not hesitate to teach that an imperialistic policy will tend to bring about wars, though I presume I must watch myself carefully if I do not want the D.A.R. after me!"[50]

While educators varied widely in their beliefs about and approaches to pedagogy, those who advocated for teaching world citizenship were generally cautious to cast the ideal as advantageous to the nation. There were a number of ways to do this. First, some educators emphasized lessons in world citizenship as a form of peace education and war prevention that was fully compatible with American patriotism, noting that internationalism and "wholesome nationalism" were "complementary," "not opposing," forces. "The development of international loyalty does not abrogate or in any way invalidate our national allegiance," explained E. Estelle Downing, the head of the Committee on International Understanding at the National Council of Teachers of English, in 1925. "The citizen of the world merely takes on a new and larger loyalty while keeping the old and lesser one. Along with the Stars and Stripes he hoists what [G. K.] Chesterton calls 'the flag of the world.'"[51] Such claims created space for educators to simultaneously support internationally themed lessons *and* American involvement in both of the world wars, casting the conflicts as a battle for safeguarding and spreading democracy as well as forging a more cooperative internationalist future. These discourses also allayed concerns, especially prevalent during the 1920s, that the promotion of international attitudes could lead the children of immigrants to reject assimilation in favor of retaining their old-world values or "hyphenated" identities.

Additionally, teachers often described a world-minded education as a process of learning to "cooperate" with different nations through increased commerce and trade. Such characterizations cast the ideal as less of a moral imperative than a transactional engagement with different countries to ensure the continued march of "progress" and economic development within the United States and around the globe. As Lester Ade, the superintendent of public instruction in the Pennsylvania schools, wrote in 1938, "Interdependence, understanding, and cooperation" were new ideals that had "been made essential by the swift development of technology, industry, and communication. Through these mediums western civilization is being merged into a new world civilization and a new obligation is imposed on all our citizens." It was a "paramount function of education" in this era to teach a "sympathetic understanding and mutual toleration among the different races, religious and cultural groups of the world . . . and to promote a new world economy which gleans benefits from all that is good in all groups." This vision of world citizenship supported the United States' continued economic ascent along with the global spread of commodity capitalism and Western values, and befitted an American curriculum that was increasingly oriented toward preparing young people for labor in an industrial society.[52]

Finally, the terms "world-mindedness" and "world citizenship" were invoked by educators not just to promote a generic, outward-looking internationalism but also to create a more palatable, reassuringly Eurocentric vision of the demographic diversity within the United States. In the 1910s and 1920s, Americans increasingly regarded their nation as a microcosm of the world—"a melting pot of nations"—and the public school as one of the most important spaces in which Americans of different backgrounds learned to cooperatively live and work alongside each other. While efforts to teach internationalism and intercultural tolerance during this time were understood as a rebuke of xenophobic nativism and rigid Americanization programs, they nevertheless overwhelmingly focused on understanding the peoples of Europe, often with only a cursory or generalized reference to the peoples of Latin America and Asia. African Americans—commonly considered by whites to be "nationless"—as well as the peoples of Africa were almost completely excluded from, or severely misrepresented as "backward" or "uncivilized" in, mainstream lessons and curricular materials on world-mindedness until the late 1930s. As Zoe Burkholder writes, such patterns "reflected the extent of white supremacy and racism in the United States that fostered a climate where teachers could openly teach acceptance of white ethnic groups, but not yet promote acceptance [of] American Indians, Asian Americans, Latinos, or African Americans." Throughout the 1920s and 1930s, white liberals could thus show devotion to teaching world-mindedness while remaining wholly "not attuned to the problem of racial discrimination in America."[53]

In addition to privileging whiteness, educational programs in internationalism and world citizenship frequently had a self-congratulatory and evangelizing impulse, suggesting that the United States, with its uniquely multicultural makeup, already possessed an inherent capacity for "tolerance" and "mutual understanding" that most other nations did not. Some educators believed with zeal that the nation could, with proper treatment in the schools, effectively scale up and export its model of liberal, enlightened, and cooperative citizenship from the classroom to the community to the world at large. As Ade reasoned in 1938, against a backdrop of rising fascism and authoritarianism in Europe,

> America has been called the melting pot of nations. Peoples from the four corners of the earth inhabit our territory. These races, divergent in their inherited and acquired characteristics, have become homogeneous since their coming to America where they have had the opportunity of mingling together. The American ideal of freedom lends itself forcibly to the development of a world citizenship. . . . In international relations, as in relations within a smaller group, the American ideal of freedom dictates the right of an individual to make his own decisions. Only such a wide sphere of individual freedom is compatible with democratic freedom. If the nations of the world are to become a harmonious neighborhood of peoples this ideal of freedom—industrial, political, religious, and cultural freedom—must be maintained. Society looks to education for support of this essential principle.[54]

Through such discourses, the idea of world citizenship gradually became neutered of any radical or reformist impulse, and found its place in a largely conservative educational system. The aim of teaching appreciation for the "brotherhood of man" was reoriented toward awarding strategic advantages to the nation in the form of modernizing the educational system for an international and industrial economy, promoting patriotic unity while ensuring white supremacy at home, and strengthening American-style democracy, markets, and influence around the world.

The term "technologies of global citizenship" is thus intended to capture the ambiguity of these ideals, highlighting how teachers marshaled together an array of sensory media and activities to simultaneously fashion an expansive, cosmopolitan "window" into the world as well as a carefully managed view of the peoples within it. The term "technologies" denotes not only the plurality of devices that educators used to teach about the world but also the regulatory logics and discourses that shaped their production, promotion, and use. In her study of American social reform movements that spanned the nineteenth and twentieth centuries, Barbara Cruikshank coined the term "technologies of citizenship" to describe the structures and discourses of liberal democracy that purport to advance individual autonomy and civic participation while belying their own restrictive governmentality. Drawing from Michel Foucault, she argues that states exert power over citizens by limiting how they are taught to govern and help themselves through voluntary as well as participatory action. These "participatory and democratic schemes," she writes, "do not cancel out the autonomy and independence of citizens but are modes of governance that work upon and through the capacities of citizens to act on their own."[55]

While technologies of global citizenship promoted an unprecedentedly outward-looking view in the classroom, these films, radio programs, pageants, and other "devices" regularly drew on the domestic and assimilationist myth of the "melting pot" to teach about the virtues of international cooperation, and position the United States as a model of multicultural harmony for the rest of the world to follow. Such devices encouraged American students to celebrate their nation's own historic capacity for intercultural "harmony" while ignoring the past and present struggles of minority groups within their shores. Indeed, in a school system heavily segregated by race, the ideal of a media-enriched education, like that of world citizenship, was at once progressive for its time and predictably exclusive. Both were educational priorities that were articulated primarily by white educators and imagined for the benefit of white students.

Mary Burnett Talbert, a prominent Black educator, suffragist, and civil rights activist, recognized by the end of World War I that the emerging ideal of world citizenship would not be extended to all Americans. As African American troops returned home from Europe, they found that despite the United States' rhetoric of "making the world safe for democracy," it continued to tolerate its own systemic oppression and disenfranchisement of Blacks at home. In an address before the Third World's Christian Citizenship Conference in Pittsburgh in 1919, Talbert described how Black Americans were being denied the "right to world citizenship" just as they were denied the right to basic citizenship in the United States. She noted how the children of European immigrants were being welcomed into the nation's public schools, and included in its universalist and expansionist rhetoric, while Black children were excluded and ignored—despite the participation and sacrifices of Black troops in the war. Talbert argued that in this moment of rising American power on the world stage, all citizens

> should feel what our country is—her position among all nations of the earth, her duties, her responsibilities, and her capacities—and how grand and important the part she is to play on the great stage of human experience for centuries to come. It is impossible for the Negro to have this feeling, if he is constantly made to feel that he is not a part of this country. We read "All welcome." We wonder what is meant by "All." "Let Everybody Come." We wonder who is "Everybody," for we know we are not included in the "All" or "Everybody."[56]

White liberal educators' interests in teaching world citizenship gradually evolved in the 1930s, with the growth of the social sciences and alarming rise of fascist ideology in Europe, to include a concern for homegrown prejudice, stereotyping, and the mistreatment of racial, ethnic, and religious minorities within their own communities. Still, most discussions of internationalism and intercultural understanding in schools remained stubbornly abstract, and removed from domestic matters of segregation and racism. The ideal of world citizenship that subsequently took root in American education was a conservative one, firmly oriented toward what Bruce Robbins describes as "morals and sentiments rather than agents and politics." The "world citizen," as Talya Zemach-Bersin contends, became an "apolitical" citizen conjured through a rhetoric of "universal kinship and belonging" that "disguises the politics and power structures that are tied to the interests of and allegiances to the nation-state."[57] For educators and reformers throughout the twentieth century, world citizens could, through well-meaning practices of mediated encounter, learning, and intercultural exchange, transcend the legacies of racism, imperialism, nationalism, and power inequalities without having to systematically grapple with them.

These historic contradictions in civic education initiatives are relevant to the emerging body of critical scholarship on global citizenship education. Here, scholars have pointed out that the relatively recent explosion of techniques designed to teach "global citizenship" in K–12 and higher education, such as the study abroad programs, pen pal exchanges, and other cultural exchanges that have been steadily gaining steam in the United States since the beginning of the Cold War, employ an altruistic rhetoric of liberal intercultural exchange, citizen diplomacy, and strengthening global understanding and democracy through "people-to-people" interactions. Placing the power on citizens to create a "better world," these technologies of global citizenship obscure their inherent governmentality, and draw attention away from the ways in which industries and states may be actively undermining intercultural relations and global peace. Global citizenship education initiatives, from the pen pal movement that emerged in American education after World War I, to the People-to-People Program launched by the Eisenhower administration during the Cold War, have brought untold educational benefits and deeply meaningful intercultural relationships and experiences to participants. But they have also been instrumental to larger US interests of national security, statecraft, and delivering a competitive advantage to the nation and its allies by equipping citizens with strategic cultural, linguistic, and technological skills and knowledge.[58]

While a small number of American students and teachers participated in international travel programs before World War II, most did not have the means to travel far from home, and made use instead of a wealth of established and newly available media technologies to acquire an international outlook on the world. My contention is that these media were more than mere AV aids; they were powerful "devices" for efficiently preparing children to understand and uphold the dominant social values of native-born European Americans within an increasingly diverse United States, and ensure a dominant role for the nation in a changing international order. Educators embraced these devices not merely as transmitters of academic information but also as participatory instruments of local, national, and global social reform that operated through the reformation of students' senses and subjectivities. Through media-enabled acts of viewing, hearing, sensing, and communicating with cultural others in and outside of the classroom, students would rehearse, at what Cruikshank calls "the microlevels of everyday life," the attitudes and habits of mind deemed necessary for building a cooperative society at home as well as advancing American involvement, industry, and influence around the world.[59]

In all, the early twentieth century saw a surge in communication practices in schools that were local in formulation but ambitiously global in tone. Harnessing a mix of new AV technologies, grassroots pageants and performances, and international exchanges of humble and homemade media, ordinary teachers and students engaged in and helped forge vibrant yet problematic practices of global media reception, cultural production, and circulation in American schools and society. In their efforts to utilize new approaches to communication to broaden students' worldview, educators and reformers aimed to refashion schools into powerful informational agencies that could rival mass media, uplift communities, strengthen democracy, and promote international understanding. Through these sorts of activities, educators believed they could strike a balance between the pull of conservative nationalism and the progressive imperative to promote peace and democracy at home as well as in the international world. As Dewey wrote in 1916, "We are now faced with the difficulty of developing the good aspect of nationalism without its evil side—of developing a nationalism which is a friend and not the foe of internationalism. Since this is a matter of ideas, of emotions, of intellectual and moral disposition and outlook, it depends for its accomplishment upon educational agencies, not upon outward machinery."[60] But despite their faith that school-based communications could counteract the divisiveness and prejudices bred by industrial modernity and mass culture, educators' own mediations of the world were often marked by the same ideological contradictions and hegemonic thinking from the dominant culture that they wished to reform.

2   Sightseeing in School: Promotions of Virtual Experience and World Citizenship from the Stereoscope to Radio

Through the loop-hole of retreat;
Peep at such a world;
Hear the great Babel;
And feel not the crowd.
— *The World Visualized for the Classroom*, 1915

In 1930, California educator Anna Verona Dorris penned a sweeping essay titled "Educating the Twentieth-Century Youth." The former president of the NEA's Department of Visual Instruction and author of an influential book, *Visual Instruction in the Public Schools* (1928), Dorris had become a leading figure in the visual education movement, a campaign by American educators, university researchers, and media producers to improve teaching and learning with the help of new media technologies and sensory aids.[1] In her essay, Dorris pressed her case for bringing more media into the classroom, calling on teachers to reject "formal and bookish" approaches to instruction, and explore the pedagogical potentials of newly available audio and visual devices, including motion pictures, lantern slides, photographs, and radio. She provided a rationale for why such tools were a good social investment, proposing that they would not only improve the efficacy of teaching in the present but also promote an "international consciousness" and "new type of citizenship" in the future.

"Change, change, everything has changed—is constantly changing the world over," she wrote. In this era of rapid change, Americans remained "woefully ignorant" of world geography and foreign cultures—a deficiency that teachers had a duty to correct in the wake of the Great War. Fortunately, she reasoned, thanks to the "wealth of visual material which this scientific age has developed for our use," teachers could take students on a virtual tour of the world, giving them vivid "pictorial experiences" that would spark a lifelong interest in world cultures and affairs:

> Through these pictorial experiences students are transported mentally to the remotest corners of the globe. They may penetrate the frozen regions of the far north, and actually live—in imagination—with the Eskimo and his interesting family. Again through this same medium they may visit India, Japan, or China and not only study firsthand the geography of the countries, but they may visit homes, schools, factories, or ancient palaces or temples. In other words, through these magic devices every student may see for himself how our foreign neighbors work and play; how they actually live day by day. Such concrete fascinating experiences could never be gained from the printed page.[2]

Like many of her contemporaries, Dorris envisioned a twentieth-century classroom filled with cutting-edge media technologies ("magic devices") that would enable students to virtually experience a wider world through the senses. Her praise for educational aids reflected, on the one hand, a sea change in American education that had been underway since the late nineteenth century, as progressive educationalists began calling for moving beyond the traditional teaching methods of verbalism and rote memorization in favor of more experiential and sensory forms of learning.[3] But her comments also testified to more recent efforts undertaken by educators, commercial media producers, and policy makers to legitimize popular media technologies—then associated with entertainment and leisure—for instructional purposes. Over the course of the visual education movement, which began in the early 1900s and dissipated at the end of World War II, when AV aids became commonplace in American schools, the notion that media could bring students into contact with distant lands and give them "vicarious experiences" of foreign or faraway phenomena was one of the most commonly invoked justifications for its uptake in education. As one educator put it in 1927, "It is obviously impossible to take the pupil to the end of the earth for original and direct experience. An alternative would be to bring the world to the pupil." Visual aids offered "some way of getting vicarious experiences about the world in order to make our teaching effective."[4]

This chapter examines the global, experiential, and civic rhetoric that shaped the early promotion and adoption of visual and audio technologies in American schools. Rooted in both nineteenth-century understandings of media as an edifying instrument of exploration and armchair travel and emergent thinking in the machine age about the socially transformative benefits of technology, this rhetoric has shown remarkable durability in facilitating the uptake of various waves of new media in schools since then.[5] Since the turn of the twentieth century, educational motion pictures, radio, and television have been touted as "magic" tools that can teach students about the complexities of the modern world by exposing them to distant places and foreign cultures.[6] A more recent manifestation of this discourse can be seen in millennial idealizations of the "global" and "wired" classroom, a high-tech, networked learning environment in which computers and the internet are described as helping teachers to "bring the world into the classroom," and students to "participate fully in their interconnected world."[7] This chapter argues that these global and experiential connotations of educational technology took root in the early twentieth century, when reformers began forging a rhetorical link—what cultural studies scholars call an "articulation"—between the use of various instructional media in schools and an emerging ideal of "world citizenship."[8] This progressive vision, ostensibly aimed at teaching US youths to become more aware of foreign affairs and appreciative of cultural differences in the midst of increasing international interdependence, simultaneously served the interests of the state by seeking to unify heterogeneous immigrant populations around common values of patriotism, American exceptionalism, industrial capitalism, and democratic citizenship.

This chapter first discusses how the cosmopolitan promotional discourses about emerging media served the interests of three groups of early twentieth-century actors—educators, commercial media producers, and policy makers—and then traces how these groups mobilized them in cooperation with each other to pave the way in schools for a constellation of new consumer technologies. In the second part, I consider three turn-of-the-century technologies of genteel domestic entertainment and private enlightenment—stereoscopes, lantern slides, and *National Geographic* photography—that became widely accepted as legitimate aids in the classroom by the late 1920s. In the third part, I consider the two ascendant technologies of interwar mass entertainment—cinema and radio—that were incorporated less widely into schools, but were nevertheless subject to the same promotional discourses about the experiential, worldly, and uplifting possibilities of new media. By working to legitimize these various devices for everyday scholastic use, the earliest boosters of mediated instruction breathed new life into long-developing cultural discourses that equated technological novelties with virtual experience, controlled inter-cultural encounter, and universal uplift.[9] At the same time, they suggested that the use of such technology in schools could impart a useful "world-mindedness" onto students that would serve the emergent political priori-ties of managing ethnic diversity at home while assuring a more prominent role for the United States in the changing international order.

"Efficient Substitutes for Experience": The Value of Virtual Travel in Early Twentieth-Century Education

While the educational possibilities of new media were of interest to a num-ber of groups in the early 1900s, commercial producers were among their earliest and most vocal promoters. From the devices already established in middle-class homes, such as the stereoscope, lantern slide, and illustrated magazine, to the more novel and controversial popular cultural technolo-gies of cinema and eventually radio, the manufacturers and distributors of America's burgeoning media industries saw an opportunity to expand their markets into schools. Their optimism was buoyed by the rapid growth of the American educational system. During the Progressive Era (1890–1920), the rise of industrial capitalism, immigration, and crowded cities prompted a series of reforms in education, including the expansion of the public school system, and efforts to transform the school into an instrument of mass education and social betterment.[10] Growing school populations put pressure on educators to economize their time in the classroom and instruct students more efficiently. At the same time, the rise of progressive educational theory, led by the pragmatist philosopher John Dewey, ush-ered in child-centered reforms and experiential teaching methods intended to make education less regimented and more engaging and democratic.[11] Together, these developments reflected a declining faith in rote verbalism and growing enthusiasm for modern teaching approaches that would fos-ter "interest learning," and provide students with more "concrete" and sensory experiences in schools. In this climate, with educators striving to make their lessons both more efficient and engaging, media producers saw a unique opportunity to peddle their wares.

As media companies began reaching out to the educational community, many attempted to align their promotional rhetoric with the goals of Progressive reformers. Echoing the uplifting discourses of social workers and educators, industry representatives delivered presentations at educational conferences, and published articles and advertisements in pedagogical journals in which they stressed a range of positive social as well as behavioral effects of using their particular audio and visual products in schools.[12] While each industry's approach to wooing educators was unique, their appeals contained many of the same themes and justification discourses. Common themes included the efficiency of technology in transmitting knowledge to learners with minimal effort; the economy of educating large numbers of students with a single device; the ability to arouse their interests through vivid images, sounds, and emotional appeals; and the power to teach desirable behaviors in hygiene, health, and morality through stories and dramatizations.

But one claim that surfaced again and again across advertisements for classroom technologies was the notion that they could serve students as sources of, or substitutes for, long-distance travel and "experience." Progressive educationalists valorized experience as the antidote to passive, verbalistic learning and a requirement for preparing youths for meaningful participation in a complex, democratic society. Writing on "Democracy in Education" in the *Elementary School Teacher* in 1904, Dewey argued that schools should foster "original" and "first-hand experiences" for the child, noting the value of field trips, nature study, and other observational activities in "widening and organizing his experience with reference to the world in which he lives."[13] For commercial producers of visual aids, making references to far-flung people and places was a way to boldly convey the experiential value of their products. When Underwood and Underwood, a leading manufacturer of lantern slides and stereographs, sought to expand its market into schools, the company published product guides for teachers that promised its visual aids would foster "real experiences of seeing distant objects and places," and "extend the environment of the schoolroom to the whole world, giving the pupils the personal experience of being in every country and actually coming into personal contact with the various industries and activities of the world." The *Catalogue of Educational Motion Picture Films*, published in 1910 by George Kleine, one of the first educational film distributors, opened with a lengthy panegyric from University of Chicago anthropologist Frederick Starr titled "The World before Your Eyes," intended to convince hesitant educators of cinema's instructional value. After naming a litany of exotic countries that he had virtually visited, Starr concluded, "No books have taught me all these wonderful things; no lecturer has pictured them; I simply dropped into a moving picture theater at various moments of leisure, and . . . I have learned more than a traveler could see at the cost of thousands of dollars and years of journey."[14]

The notion that visual technologies could stand in for firsthand experience of worldly encounters is traceable to Enlightenment era practices of colonial exploration and scientific exhibition. As Barbara Stafford argues, modern visual education is predicated on the explosion of visual devices—such as cabinets of curiosities and magic lanterns—that visualized, condensed, and organized the world for the Western subject in the eighteenth century. As Europe extended its imperial reach across the globe, scientists and explorers published popular illustrated accounts of their expeditions, while producers of optical devices and didactic children's toys used exotic imagery to demonstrate the power of their products to overcome the barriers of time and space. Bourgeois citizens in Europe and the early United States used these devices to educate themselves and their families

Figure 2.1
Advertisement for Pathéscope Company of America. *Reel and Slide* 2, no. 5 (1919): 34.

on geography, history, and the sciences; entertain themselves; and display their worldliness and education for others.[15] In the public sphere, global visual attractions later emerged in the form of natural history museums, panoramas, and international expositions, offering what Richard Altick calls the "bourgeois public's substitute for the Grand Tour."[16] Through the visual consumption of foreign landscapes, oddities, and exoticized bodies, Westerners advanced the projects of nationalism and imperialism by rehearsing the racial hierarchies and narratives of technological progress that placed white Europeans at the center of civilization.[17]

In the late nineteenth century, the rise of industrial production led to rapid advances in visual technology, including photography, stereography, and lantern slide projection, and brought these devices within the reach of a growing middle class. In the postbellum United States, chautauqua lecturers, world's fair exhibitors, missionaries, and itinerant showmen took up lantern slide projectors and early motion picture technology to adapt the global visualizing traditions of the Enlightenment for mass consumption and popular instruction. These attractions became important conduits of cultural capital and citizenship formation. As X. Theodore Barber notes of the illustrated travel lectures, or travelogues, that gained popularity during this time, the "heightened sense of culture and refinement surrounding [these] exhibitions attracted the 'better classes' as well as those who wished to be identified with them."[18] Worldly visual media, such as the oleograph set of two hundred stereoscopic views sold by Sears, Roebuck and Co. for use in the home and school, mixed touristic images of foreign lands with symbols of American expansionism and power, including national parks, government buildings, industrial and military structures, and the newly acquired territories of Alaska, Hawaii, and the Philippines. For both native-born white Americans and the immigrant audiences seeking upward mobility as well as acceptance in their new land, this global mass visual culture served a dual purpose: it broadened spectators' cultural horizons and sense of refinement while it interpellated them as a unified public of American subjects.[19]

The capacity of new visual media to simultaneously teach cosmopolitanism and conformity to the masses made it appealing to social reformers at the turn of the twentieth century. Facing the concurrent expansion of American empire abroad and an influx of immigrants at home, political leaders struggled to address new and competing priorities of awakening

Americans' interest in the international world while consolidating their sense of national identity and loyalty to the United States. As conservative and nativist groups called on schools to teach "100 per cent Americanism," defined as patriotic, English-speaking participation in mainstream cultural and civic life, liberals urged teaching "tolerance" and "world citizenship," a generalized, outward-looking appreciation for other nations and cultures as a basis for promoting interethnic and international cooperation.[20] After World War I, as the United States assumed a more interventionist role in international affairs, progressives and peace advocates managed to sustain the ideal of teaching world citizenship in public education amid rising patriotic fervor by casting it as a matter of both domestic and global security that would enhance, not erode, one's national loyalty.[21] As Isaac Kandel of Columbia University Teachers College wrote in a 1925 essay titled "International Understanding and the Schools," "An international attitude can in no way involve a repudiation of patriotism; indeed, because its basis should be a recognition of the part played by other countries as well as our own in the progress of the world, it should intensify patriotism and quicken the consciousness of those things that make our own country great in the service of humanity." Many educators believed that if immigrants could be taught to cast aside their Old World loyalties and native-born Americans their provincialism, a more unified generation of patriotic yet world-minded citizens would emerge to prevent, or at least be better prepared for, future wars. As former president William Howard Taft wrote in the introduction to *A Course in Citizenship and Patriotism*, a wartime curricular guide produced by the American School Peace League (ASPL), a peace education organization, the war presented schools with a dual obligation to not only transform America's immigrants into "law-abiding, patriotic citizens [who will contribute] . . . prudential virtues, and civic activity, to the general welfare" but also "impress upon the youth . . . the idea that we are not the only people in the world; that we should earnestly cultivate friendship and sympathy with other peoples."[22]

While historians have noted how these diverging definitions of citizenship were taught in schools through curricular reforms and textbooks, they were also disseminated by the era's new media.[23] Indeed, a core assumption that fueled the visual education movement's rise in the 1910s and 1920s was that visual and sensory aids were better suited to teaching the complexities of twentieth-century citizenship than traditional, "dry" textbooks

and verbalism. Media manufacturers and educators alike touted visual technologies as more "vivid," "vital," and capable of leaving "impressions" on
young people's malleable minds than the written or spoken word, making
them particularly useful for familiarizing students with the changing world
within and beyond their shores.[24] Travel-themed media in the form of stereographs, lantern slides, motion pictures, and radio became particularly
apt vessels for the contradictions of citizenship education in this period.
Their capacity to represent of a mixture of national and international sights
and sounds meant that they could be easily mobilized to teach both the
inward-looking values of patriotism, nationalism, and "industrial citizenship," which emphasized the formation of workers for an industrial economy, and the outward-looking ideal of "world citizenship," which stressed
greater cooperation among nations through cultural exchange, diplomacy, and trade.[25] These ambivalent interests can be seen in the frequency
with which references to the Panama Canal appeared in advertisements
for visual education.[26] A critical opening in international trade completed
and controlled by the United States in 1914, the Panama Canal—and the
widely reproduced claim that children could virtually "see" or "visit" it

Copyright Keystone View Co.

She *Sees* the Panama Canal

Figure 2.2
Advertisement for Keystone View Company. *Visual Education* 1, no. 6 (1920): 63.

without leaving the classroom with the help of lantern slides, stereoscopes, or films—emblematized a broader educational interest in familiarizing students with not only the United States' industrial prowess but its increasingly powerful role in shaping global trade and affairs too.

Early promotional discourses for educational technology were thus deeply contradictory, suggesting that these products could both expand the pupil's worldview and control it in accordance with dominantly held values. These contradictions were on display in a series of 1928 essays in the *Educational Screen* by Ercel McAteer. In the first, titled "The Influence of Motion Pictures upon the Development of International Co-operation," McAteer argued that educational films could promote world peace by erasing prejudice, "placing before the children of all nations knowledge which will create in them a sympathetic understanding of their brothers in foreign lands." In stark contrast, her second essay, "The Influence of Motion Pictures in Counteracting Un-Americanism," contended that the same technology should be taken up to teach Americanization, promote patriotism, and counteract the "repulsive creature" of Bolshevism by showcasing the merits of a free market system built on "self-reliance" and "individual incentive."[27] Through discourses like these, visual education advocates suggested that the cosmopolitan subjectivities fostered by new media would be of a generalized variety—emphasizing universalism, or the belief that people are "fundamentally the same the world over"—and would not interfere with students' development of mainstream American identity and ideals. In this way, incorporating new media into education was a matter of more than merely enhancing learning. Rather, it was a question of preserving the dominant vision of American identity and democracy in an increasingly heterogeneous society and interconnected world.

From the Parlor to the Classroom

Stereoscopes and Lantern Slides

The uplifting claims that surrounded promotions of early classroom technologies would have sounded familiar to much of the American public in the first decades of the twentieth century. Similar discourses had been mobilized for years in advertisements for stereoscopes and lantern slides. These devices' popular associations with genteel amusement and armchair travel in the late nineteenth century made them well primed for conversion into technologies of mass instruction in the early 1900s.[28]

Developed in the 1860s by the medical professor and writer Oliver Wendell Holmes Sr., the stereoscope was a handheld binocular device designed to display stereograph photography, also known as stereograms or stereo views, in three-dimensional detail. Though stereographs were produced on a multitude of subjects, the most popular varieties focused on travel and sites of cultural or historical interest, including Greece and Rome, colonial territories, the American West, and the Holy Land. As Tom Gunning observes, the rise of stereoscopy as an American pastime was indicative of a growing fascination with "foreign views" in popular culture at the turn of the century. The theme of visual consumption as virtual travel was salient in a range of visual and technological attractions in this period, including amusement park rides, world's fair expositions, museum exhibits, illustrated lectures, and early travel films. Each offered the public a "means of appropriating some distant place through an image, of seeming to be somewhere by being absorbed in a 'view.'"[29]

Stereograph manufacturers brought foreign views into the home in the 1890s with the release of boxed sets of travel images advertised as virtual "tours." These included the *Trip around the World* set by Underwood and Underwood, *100 Views of the World* by Sears, Roebuck and Co., and the *Tour of the World 400 Series* by the Keystone View Company. In 1900, Underwood and Underwood trained its traveling salesmen to recite the following script while canvasing residential neighborhoods, selling the *Trip around the World* set as a substitute for travel, source of social distinction, and tool for giving young children a leg up on their studies before entering school:

> With our "Trip" you, your wife, children, and friends can make this great journey right by your own fireside, and obtain truer ideas of the famous scenes and the life and customs of the people in these distant countries than you could obtain any other way. . . . The "Trip" affords very impressive Kindergarten instruction to your children of places they will constantly be studying and reading about.[30]

Many of the same companies that produced stereographs also made glass lantern slides and eagerly marketed both products side by side as essential equipment for the modern school. Like the stereograph, the photographic lantern slide gained popularity in American culture in the late nineteenth century, and was heavily associated with armchair travel and popular instruction. At the turn of the century, Americans commonly encountered the technology in the form of commercial travel exhibitions and illustrated lectures.[31] Known as "travelogues"—a term coined by the lecturer Burton Holmes—these popular theatrical entertainments were performed by touring showmen who screened images of foreign scenery and people, along with spoken commentary, sound effects, props, and even costumes, to take audiences "around the world in one evening." In small towns and rural areas less frequented by lecturers, citizens could stage their own travelogues using mail-order sets of slides and prefabricated scripts. Until the 1930s when they were eclipsed by travel cinema, illustrated travel lectures drew large audiences, and blurred the lines between mass education and entertainment. They offered a refined and respectable form of public diversion for a growing middle class interested in learning about the world and experiencing, even if only through imagination, greater geographic mobility.[32]

By the early 1900s, producers of stereograph and lantern slides began to focus on getting their products into schools. In 1907, the Keystone View Company of Meadville, Pennsylvania, published a handbook to assist teachers in designing lessons around its new "600 Set" of travel views, which it manufactured in both stereograph and lantern slide format, and packaged in a large wooden storage cabinet. An elaboration on the concept of the boxed "world tour," the 600 Set claimed to contain images of every US state and country in the world. The company worked with educational experts to cross-reference the images by academic subject so that each image could potentially be applied to a variety of lessons. As Meredith Bak observes, this "almost encyclopedic organization" of the 600 Set endowed the product with progressive ideals of rational efficiency and likely helped to allay educators' concerns about using a device associated with domestic leisure for teaching purposes.[33] Citing its usefulness in teaching a more world-minded history and civics, a Massachusetts normal school educator wrote in the teacher's guide that the "striking scenes . . . will do much to help pupils to *think* the events of history into their correct geographic setting," thus making "real and vivid a large body of facts of vital importance to every citizen."[34] Presenting a comprehensive collection of images of the world in a sturdy piece of media furniture, Keystone portrayed the 600 Set as a modern-day cabinet of curiosities and authoritative reference tool with limitless recombinant potential.

But repackaging old touristic views for use in the schoolroom resulted in visual lessons that were often ill suited to the academic needs of teachers, and more in line with the pleasurable scopic conventions of popular stereography and travelogues.[35] As Mark Jefferson, a geography professor and early adopter of stereographs at the State Normal College in Ypsilanti, Michigan, complained, commercially prepared "pictures have been taken from a scenic rather than a geographic point of view." In his estimation, teachers were better off buying a stereoscopic camera and taking geographic photographs of their own to create visual aids more suited to their lessons.[36]

Despite the industry's efforts to tailor its products to the needs of teachers, its advertisements in educational journals largely recycled popular tropes of leisured armchair tourism, foregrounding the products' dazzling verisimilitude over their educational content or academic applications. Underwood and Underwood and the Keystone View Company frequently displayed both their lantern slide and stereograph equipment in the act of visualizing images of people and scenes in Latin America, Africa, and the Orient, drawing attention to the technologies' power to hurdle the barriers of time and space. Reminding educators that "learning comes through experience," Keystone claimed in one of its advertisements—featuring women hulling rice in the Philippines—that its images were "efficient substitutes for first-hand experience."[37] As Keystone repeatedly asserted, the stereoscope's immersive views were so realistic that some pupils could not help but instinctively step backward after viewing an image of a deep gorge.[38] Similarly, Underwood and Underwood's handbook for teachers, *The World Visualized for the Classroom*, stated that the stereoscope could momentarily overtake the spectator's consciousness, replacing it with "real experiences of seeing distant objects and places":

> It is a scientific fact that while looking through the instrument, it is not only possible but it is easy and natural for one to lose all consciousness of immediate bodily surroundings and to gain *real experience of seeing*, of being *present in* the places themselves. . . . [T]he stereoscope annihilates time and distance and transports us in the twinkling of an eye to the very heart of the distant scene.[39]

Throughout the interwar period, many educators who published their own accounts of teaching with lantern slides and stereographs echoed this global and experiential rhetoric from the industry, describing students as "transported in imagination" to foreign countries and even hypnotized into a trancelike state of wonder. James Emery, the Rhode Island school principal mentioned in the previous chapter, published a series of essays in educational journals outlining how teachers could take "the slide route" to faraway lands, including Africa, India, Japan, and Latin America, suggesting that "it might be possible for the slide user to approach every continent and every country by the same hard-and-fast method."[40] The commercial travelogue genre was so popular that some school administrators worried it prevented teachers from designing lantern slide lectures that were sufficiently academic. Speaking at the inaugural meeting of the NEA's Committee on Visual Instruction in 1916, A. W. Abrams, chief of the newly formed Visual Instruction Division of the New York Department of Education, warned educators about the "canned lectures" and "fixt sets" of geography slides that passed as legitimate visual aids in many schools. He argued that they amounted to little more than "general purpose entertainments that have no special relation to each other and no serious educational ends."[41]

Despite their claims to accurate and encyclopedic coverage of the world, commercial collections of lantern slides and stereographs presented a deeply reductive view of race, gender, class, and cultural diversity. Their ideal users, as depicted in advertisements and educational literature, were exclusively white, middle-class pupils. While some visual education materials stated that children of both sexes should learn to operate projection equipment in school, images and accounts from the era suggest that this responsibility typically fell to boys, who were presumed to be more apt with technology. This mimicked the gendered conventions of the theatrical travelogue, in which lecturers, like the expeditionary explorers they channeled, were almost exclusively male. When not gazing at industrial or natural wonders, the homogeneously white student users of lantern slides and stereographs were often depicted as looking at nonwhite peoples of colonized lands. Advertisements thus sought to underscore the technologies' visualizing authority by aligning them with the imperialist and touristic gaze typical of other visual attractions of the era, such as world's fairs, museum exhibits, and travel cinema. Representing the Western subject as the agent of seeing, knowing, and learning, this mode of mediated looking rendered the distant, non-Western, "low-tech" other as an oddity to be contained, scrutinized, objectified, and consumed by the civilized viewer.[42]

Lantern slides and stereographs reinforced racial hierarchies not just through their individual content but also through their assemblage into

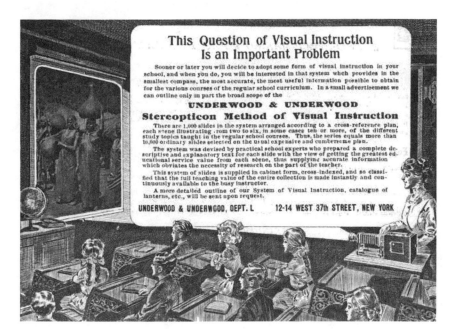

Figure 2.3
Advertisement for Underwood and Underwood. *Journal of Education* 78, no. 22 (1913): 591.

instructional sequences of images. In the Keystone View Company's 1917 guide to the 600 Set, a sample lesson on "Racial Geography" advised teachers to arrange their stereograph and slide images of the world's races in order of "dominance," proceeding from white to "yellow" to black. This chromatic ordering was intended to help students understand the hierarchies of "pure" and "developed" races, the "not so far advanced," and the "backward" and "primitive."[43] Its appearance in a mainstream visual education text highlights how easily products that claimed to "bring the world" to children could also disseminate openly racist and eugenicist views.

Makers of lantern slides and stereographs were eager to demonstrate their capacity to not only expose American students to different cultures but make students of different cultures more American. After World War I, Underwood and Underwood promoted two flagship products side by side: *The World Visualized for the Classroom*, a teacher's manual offering "1000 travel studies" using the company's slides and stereographs, and the *Underwood-Dixon Series on Americanization*, a special collection of images emphasizing citizenship education, patriotism, and the assimilation of immigrants. While the former promised to "connect the schoolroom to the world outside" through a comprehensive representation of nations, races, and sites of international interest, the latter pledged to teach "patriotic citizenship and love of country" through dramatic renderings of events in American history, including the Monroe Doctrine, Civil War, and Great War. The parallel marketing of these cosmopolitan and nationalistic visual products in educational journals in the 1920s highlights both the centrality of virtual experience in the promotion of early educational technology and contradictions of citizenship education during this period. Amid the United States' shift from an isolationist to a more internationalist foreign policy, a rise in nativist and anti-Communist fervor at home, contributing to the passage of the 1924 Immigration Act that sharply curtailed immigration, created an educational landscape in which lessons in pluralism and patriotism, intercultural tolerance and rigid assimilation, often shared space and status in the curriculum.[44]

National Geographic School Service

After World War I, a concern voiced widely by educators and social commentators was the general "geographic ignorance" of Americans—a problem believed to be worsened by the average citizen's lack of interest in international matters along with the explosion of popular media disseminating erroneous and exaggerated depictions of foreign people and places. The journalist Walter Lippmann, in his influential *Public Opinion* in 1922, introduced Americans to the concept of the "stereotype," or the discrepancy between the complexity of the "world outside and the pictures in our heads." Lippmann warned that the average citizen's growing dependence on media-manufactured stereotypes to make sense of the world posed a threat to democracy.[45] For educators interested in raising a liberal and world-minded citizenry, promoting greater international awareness and information literacy in school became urgent and mutually reinforcing priorities.

No form of media was better positioned to benefit from these concerns than the *National Geographic* magazine, which balanced the high-minded respectability of a preeminent scientific organization with the popular appeal of travel photography. In 1919, the National Geographic Society launched a school service, the Pictorial Geography Series, and appointed its creator, a geography teacher named Jessie Burrall, to promote it to educators as a "new phase of carrying out [the society's] purpose in the increase and diffusion of geographic knowledge."[46] For a small fee, teachers of history and geography could order sets of glossy reprints of images from the magazine, mounted on large cards for students to individually handle, study, and arrange on the chalk rail of their classrooms. With support from the US Bureau of Education, the National Geographic Society later published the *Geographic News Bulletin*, a weekly scholastic reader richly illustrated with photographs from its expeditions, and distributed it to seventy thousand teachers throughout the country within its first two years.

The Pictorial Geography Series extended the tendency of the *National Geographic* magazine to eschew controversial topics, and promote a view of the world that was at once cosmopolitan and Amerocentric.[47] As Catherine Lutz and Jane Collins note, through its celebration of the technology of

Figure 2.4
An image from the National Geographic Society's Pictorial Geography Series, "Benguet Brides of Luzon, Philippine Islands." From Jessie L. Burrall, "Sight-Seeing in School: Taking Twenty Million Children on a Picture Tour of the World," *National Geographic* 35 (1919): 503.

photography and portrayals of the world's "backward peoples" as peacefully
evolving toward modernity, the magazine "reinforced America's vision of its
newly ascendant place in the world by showing how far we've come."[48] This
spirit permeated the use of the Pictorial Geography Series in educational
settings. The National Geographic Society touted it as a tool for both geography and Americanization instruction, beginning in 1918 with a trial run
as a device for educating foreign-language-speaking soldiers in a US military
training camp in California.[49] At the war's close, Burrall delivered speeches
before the NEA and published essays in teaching journals to inform her
fellow educators that teaching geography with *National Geographic* pictures
was necessary for promoting "world unity" while making "true Americans
of all boys and girls." By placing its global images and bulletins in schools,
she argued, the society would help to fulfill the twin imperatives of postwar
education: transforming children into geographically literate "world citizens" while providing a "sound basis for all patriotism":

> We are now almost automatically a world power, and every boy and girl in our
> public schools must be prepared to be not only a worthy citizen of the United
> States, but also an intelligent and helpful member of a world family. . . . That can
> only be promoted through a sympathetic understanding of world peoples, and as
> long as time and expense limit extensive travel, the necessary knowledge of world
> peoples can only be secured through a study of innumerable pictures.

For Burrall, like many other boosters of educational technology after World
War I, outfitting classrooms with vivid, up-to-date visual aids was important not just for the immediate task of geography instruction but also for the
work of preparing American youths to assume the role of a "world power"
in a more internationalized future. The pictures would also unite America's diverse students in a vision of twentieth-century citizenship defined by
industrial capitalism and the global flow of commodities—sparking their
interest, for example, in the faraway origins of bananas displayed at the corner grocery—and promoting engagement with current events by "bringing
maps to life" and making them "glow with fire and meaning."[50]

The US Bureau of Education echoed these claims in its own promotions
of the *Geographic News Bulletin* after its launch, noting that "to enable the
school children of to-day to perform their duties as citizens of to-morrow
they should have a knowledge of other peoples and nations."[51] *Visual Education* magazine agreed that the school service offered not merely a useful
aid to the academic study of geography but a boon for the United States as

well in its cultivation of a new kind of citizenship that was at once world aware and patriotic: "This 'seeing' of other countries than his own in an intimate and concrete way is the child's substitute for travel and so tends to build the foundation for a genuine understanding of international problems and is one of the most effective instruments for bringing about true Americanization."[52] Through these discourses emphasizing an ambiguous mix of strategic cosmopolitanism and Amerocentric conformity, educators and policy makers reinforced the promotional rhetoric of the burgeoning educational media industry. Together, they underscored the adaptability of established media of virtual travel to the new century's educational tasks of combating geographic illiteracy, teaching patriotism and free market capitalism, and promoting a generalized awareness of other nations and cultures. Making room for popular devices of armchair travel in the classroom, they collectively recast them as substitutes for experience, stimulants of good citizenship, and reliable tools for teaching young Americans to embrace their country's exceptional and rising status in the new global order.

The World Is Yours: Legitimating Movies and Radio as Devices of Virtual Experience

While educators widely accepted photographic lantern slides, stereoscopes, and magazines as legitimate aids to instruction by the 1920s, their adoption of the newer and more controversial technologies of mass entertainment—motion pictures and radio—was neither as widespread nor as swift.[53] The interwar years (1918–1939) saw the maturation of the Hollywood studio system and ascendance of cinema as an integral part of mainstream American culture. It also saw the transition of radio from a chaotic soundscape of independent, educational, and amateur broadcasters to a commercial system dominated by a few corporate networks and professional programming. The growing presence and persuasive power of film and radio in American society prompted a mix of anxiety and enthusiasm among educators, who variously fretted and fantasized about their effects on children's learning, social attitudes, and behaviors. Even as a number of media-minded teachers and school systems began to experiment with using motion pictures and broadcasting in the classroom, their worth as technologies of instruction remained contested until their widespread mobilization for civilian and military purposes during World War II.[54]

Seeking to eliminate the schools' "lag" behind the general public in adopting these new technologies, educational and commercial advocates of film and radio leaned on established rhetorics of virtual travel and civic uplift to endow them with respectability as well as instructional value. Films and radio programs about travel and foreign people and places, often laden with racial and ethnic stereotypes, were singled out by educators, producers, and government officials as shining examples of how these technologies could teach "international understanding" and prevent future wars. For the nascent film and radio industries, foregrounding these world-revealing, universal, and educational aspects of their products aligned with larger strategies to not only build a market for their products in the nation's schools but avoid government regulation too. Touting their own concern for cultivating a world-minded, tolerant public, the film and radio industries honed a distinctively global and techno-utopian rhetoric that would similarly serve telecommunications and internet companies seeking to expand into schools at the end of the century.[55]

Motion Pictures

Before World War I, educators widely regarded movies as a cheap and socially corrosive amusement—a view shaped both by their controversial content and the insalubrious contexts, such as nickelodeons, in which they were shown.[56] In the early 1910s, the motion picture industry began reaching out to educators and social reformers in an effort to "sever cinema's associations with vice" as well as demonstrate its value as an instrument of social betterment.[57] In addition to promoting various genres of films in theaters as "educational," distributors took films that had run their course and recirculated them to schools, universities, YMCAs, and churches. A variety of nonfiction silent film genres, including science, sports, and newsreel films, were distributed for school use, but films about travel were among the most marketed to and welcomed by teachers. According to one contributor to the *Educational Film Magazine* in 1919, travel films were particularly valuable for school use because they endowed the subjects of geography and history with a sense of adventure and wonder. "Here the real appeal is not wrought by the presentation of foreign lands *ipso factor* [*sic*], but has its mainspring in the sense of mystery, shrouding supposed possibilities of adventure."[58] As Jennifer Peterson has shown, the profusion of travel films in the 1910s and 1920s played an important role in cinema's transition from the nickelodeon to the theater and classroom. Providing "instructive entertainment," their combination of scopic pleasures and edifying content helped elevate cinema's reputation from a technology of attraction and amusement to one of respectability and instruction.[59]

The educational and world-revealing potentials of cinema became more apparent to educators during World War I. It was the first war that large numbers of Americans experienced through film, from the newsreels and narrative films shown in theaters, to government films screened in YMCAs and schools, to the recruitment, training, and morale films used by the armed forces. The Committee on Public Information, the propaganda agency formed by the Wilson administration to rally public opinion in support of the war, commissioned private film studios and the Photographic Division of the US Signal Corps to produce films that "advertised America" to audiences in the United States and abroad. According to George Creel, the committee's head, these films, which highlighted the virtues of democracy and heroism of the American and allied forces at home as well as on the battlefront, carried "the gospel of Americanism to every corner of the globe" and "played a vital part in the world-fight for public opinion": "A steady output, ranging from one-reel subjects to seven-reel features, covered every detail of American life, endeavor, and purpose, carried the call of the country to every community in the land, and then, captioned in all the various languages, went over the seas to inform and enthuse the peoples of Allied and neutral nations."[60] In addition to films, the Committee on Public Information disseminated other forms of prowar propaganda in movie theaters, including posters, sales of liberty bonds, and speeches, delivered between the changing of movie reels, by "four minute men." To some observers in the educational community, this rapid mobilization of cinema to win the war revealed the technology's potential as a tool of mass instruction and civic mobilization. The industry's willingness to work with the government earned it sufficient goodwill to evade antitrust legislation in 1918 and expand rapidly thereafter into foreign markets, quickly becoming the dominant global producer of films. For years after the war, educators would lament that the industry and government failed to cooperate to develop films for educational purposes as efficiently as they had done during the war.[61]

Another significant development of the 1910s was that two of America's foremost industrialists—Thomas Edison and Henry Ford—began taking their own steps to get films into classrooms. Together, they helped popularize the notion that incorporating the latest media technology into schools was necessary for adapting the American educational system to an industrial and international age. Edison, known as the "Wizard of Menlo Park," was famous for his inventions in electricity, recorded sound, and moving pictures, and was a pioneering producer of short nonfiction films in the 1890s and early 1900s. He introduced one of the first projectors for home and school use, the Home Kinetoscope, along with an accompanying catalog of short educational films in the early 1910s.[62] Sharing testimonials from educators and school men, advertisements for the Home Kinetoscope touted its ability to efficiently approximate the experience of seeing the whole world firsthand, claiming that it made it "possible to bring the whole world in review and to show the places of interest, natural scenery, and the great industries in a way that falls little short of actual observation."[63] Though his film operations were soon eclipsed by the rising narrative film industry, Edison remained a vocal advocate for incorporating movies into instruction well into the 1920s, asserting that the technology would increase children's learning and contribute to America's greatness. "The best schoolhouse is the screen, the best teacher is the film," he declared in 1919. In the schools of the future, he predicted, "human teachers will be needed only to help guide and direct the minds of the pupils, but the pictures will do the instructing." Convinced that motion pictures would supplant the textbook, he likened the film to the newspaper as an up-to-the-minute instrument of mass information, nation building, and civic uplift that was fueling America's transformation into a world power: "The press and the screen together are making America great and powerful, and they will continue to make her even greater and more powerful as they remove the curse of illiteracy and class warfare and national vices and bestow upon her people the blessings of a liberal education."[64] Edison's lofty predictions were received with skepticism by most educators, but they succeeded in setting the tone for a now-familiar, industry-driven discourse about the urgency of incorporating new media technologies into education. Nelson Greene, the editor of the *Educational Screen*, blamed Edison in 1926 for generating a "vast amount of muddled argument, declaration, and prophecy . . . regarding the possibilities of films in the business of serious education," noting that his "dictum has been properly laughed at by hundreds of educators, but taken as pedagogic gospel by the thousands."[65]

While Edison touted the ability of educational film to shore up American power, his personal friend, Ford, the founder and president of Ford Motor Company, embarked on his own venture to produce films for educational purposes. In 1914, Ford established an in-house film studio, the Ford Motion Picture Laboratories, near his plant in Highland Park, Michigan. Between 1916 and 1921, the laboratories produced dozens of short films for distribution to theaters. These included subject films on travel, industry, history, and geography in a series called *Ford Educational Weekly*. Later came the *Ford Educational Library* (1920–1925), a collection of edited films and text guides produced specifically for educational settings, approved by a panel of educators, on subjects ranging from civics to agriculture, industry, geography, health, history, and nature study. The company marketed the films to schools, universities, churches, and YMCAs with the goal of establishing permanent collections across the country that could be rented out to local communities. For some school officials, the *Ford Educational Library* was the closest thing available to an organized, authoritative "reference library" of educational films.[66]

The Ford Motion Picture Laboratories were part of an array of experimental programs in education and social welfare that Ford developed at his company to create a more productive and compliant industrial workforce.[67] A proponent of Americanizing immigrants and a virulent anti-Semite, it is likely that one of Ford's motives for venturing into filmmaking was to counter what he saw as the Jewish subversion of American values through the Hollywood film industry. As a Ford representative wrote in *Reel and Slide* in 1918, "Our activities in the adoption of the motion picture screen are, in the main, the ideas of Mr. Ford himself, who has long realized the power of the screen to educate and direct the youthful mind along lines that are right, or, lines that are wrong, according to the nature of the picture and the intent of its producer."[68] The Ford Motion Picture Laboratories were set up in the same building as the Henry Ford Trade School, a vocational school for adolescent boys that was an outgrowth of the company's English School, which provided lessons in English, civics, etiquette, and personal finance to foreign-born workers.[69] As Lee Grieveson has noted, many of Ford's educational films focused on civics and industrial processes, advancing an ideal of "industrial citizenship" that emphasized the assimilation of immigrants and formation of workers for a capitalist society. But the films also aligned with the burgeoning educational interest in teaching world citizenship, encouraging students to become acquainted with the international world through images of geographic, racial, and ethnic difference. In advertisements for the *Ford Educational Weekly* appearing in educational journals, the company variously hailed the power of its films to help teachers "move the world into the classroom," versing them on global geography, and to "Americanize" their pupils.

Educators echoed the assessment that Ford films could expand students' appreciation of the international world while instilling American values in immigrants. "The activity of Ford's camera man has brought to hundreds of thousands of people a condensed picture of the natural beauty of America and the picturesqueness of many foreign customs," reported *Sierra Educational News*, the publication of the California Teachers' Association, in 1920:

> We have seen the slow Mexican peasant going to market with his laden ox-cart; and we have come to know him as a human being like our own farmers. . . .

Figure 2.5
Advertisement for the *Ford Educational Weekly*. *Normal Instructor and Primary Plans* (December 1919): 62.

[Other films] have told the story of American industry. . . . He has told the story of the making of the newspaper, of steel, of woolen clothes, of paper. He has brought the salmon industry of America's northwest to every corner of the continent, and he has brought the quaint historic remains of old towns which have not changed since the days of Washington to the millions whose knowledge of housing consists of an acquaintance with the New York subway and cramped flat-houses.[70]

But for all their claims of broadening viewers' horizons, the Ford films trafficked in many of the same racial and ethnic stereotypes that predominated in American popular culture. A film on the history and culture of New Orleans, for example, features an elderly "mammy" figure and a group of African American men playfully wrestling, with intertitles describing them as "happy as in the olden days" of slavery. Another film on the city of Los Angeles lingers on the "curious and odd" qualities of the city's Chinatown. And a popular title, *Democracy in Education*, glorifies colonial era Anglo-Americans as a "sturdy race of resourceful, independent, clear-thinking men" and the United States as a "modern Greece, giving unselfishly a full

"Americanization"
—the Teacher's New Task

The hope of America lies in the prompt Americanization of the youth of the land. Can it be done—with the children of foreign-born parents running into the millions? Yes—*It can, and it must!*

Figure 2.6
Advertisement for the *Ford Educational Weekly*. *Sierra Educational News* 26, no. 1 (1920): 50.

share of its attainments to the world."[71] This film was one of several titles from the *Ford Educational Library* to be screened at the 1922 NEA meeting, where visual education was a central topic of discussion among attendees. In all, Edison's and Ford's ventures in educational film highlight the long and problematic American tradition of industrialists and technology gurus promoting their wares and ideologies to educators, often appealing to discourses of civic and industrial development, global knowledge, and national dominance to do so.

Will Hays, another industry leader, also attended the 1922 NEA conference in hopes of winning the favor of educators. A prominent Indiana Republican, Hays had recently been appointed president of the Motion Picture Producers and Distributors of America (MPPDA) to help the film industry evade government censorship and clean up its image after a string of star scandals. Recognizing that cultivating the favor of educators was crucial to legitimizing the industry, one of Hays's first actions on behalf of the MPPDA was to speak before the NEA. In his address, he pledged that the industry would work "to establish and maintain the highest possible moral and artistic standards in motion picture production," and "develop the educational as well as the entertainment value and the general usefulness of the industry." Deploying a utopian and universalist rhetoric of technology, Hays maintained that cinema was already promoting understanding among the diverse peoples of the United States and world:

> The motion picture has carried the silent call for virtue, honesty, ambition, hope, love of country and of home to audiences speaking twenty different languages but all understanding the universal language of pictures. There may be fifty languages spoken in this country, but the picture of a mother is the same in every language. It has brought to narrow lives a knowledge of the wide, wide world. It has been the benefactor of uncounted millions.[72]

Through the additional work of Hays and other public relations figures in the mid-1920s and 1930s, the film industry forged partnerships with the educational community in order to promote instructional uses of film while ensuring the unfettered expansion of the theatrical movie market.[73] After Hays's 1922 speech, his office provided a grant of $5,000 to the NEA to appoint a committee of educators to assess the state of motion pictures in education. Headed up by Charles Judd, an educational psychologist at the University of Chicago, the "Judd Committee" reviewed theatrical films for teaching purposes and conducted the first large-scale survey of instructional film use in schools. Judd subsequently resigned from the committee, noting that it was becoming "besieged with promoters" and commercial interests. But it remained active with funding from the MPPDA until 1927, establishing a pattern of cooperation between media industries and educational groups that has defined the development of instructional technology ever since.[74]

A recurring theme in the film industry's public relations discourse in the 1920s and 1930s was the almost-magical ability of film to dismantle ignorance and prejudice by vividly exposing audiences to different customs and cultures. In this way, the industry implied, films could contribute to the twin educational goals of progressive reformers: promoting the harmonious assimilation of immigrants into American life and peace among the nations of the world. In a speech in 1926, Hays commemorated the launch of an Americanization movie program aboard the SS *Leviathan*, in which new immigrant arrivals to the United States were shown "patriotic and historical films," such as *The Life of Abraham Lincoln* and a short film called *Immigration*. He said, "The films will say to those future citizens, 'Here is America. See what America, your new home, is like. Look at me and love America.' There are three great purposes of motion pictures. One is to entertain, another to instruct, and a third to bring about better understanding between men and men and between nations and nations."[75] The motion picture industry called attention to the wide-ranging uses of film outside theaters—in factories, churches, YMCAs, and assembly halls—to promote both American unity and "international amity."[76] The director of public relations for the Association of Motion Picture Producers, Colonel Jason Joy, echoed these themes to educators in an essay contributed to *World Friendship*, a curricular guidebook produced by the Los Angeles City schools for teaching international understanding to children. "Only one instrument has been found which is universal in its scope, absolutely international in its sweep; and that instrument is in the motion picture," he wrote in 1928. "The motion picture speaks the tongue of every person who sees it. It can be packed in a tin can and shipped to any place in the world, with the full assurance that every man, woman and child who sees it will receive a definite impression not unlike the impression made on every other person viewing it." Both Hays and Joy appealed to educators by citing not only the film's ability to teach new Americans about their new home but also teach the whole world about America through Hollywood's increasingly international production practices and the rising "world consumption" of movies. "The pictures made in this country are not all what one would call American-scene pictures by any means," Joy noted. "The industry ransacks the world for stories, for directors, for actors, and for artisans. Hollywood is making pictures for the world and will continue to do so. Thus the motion picture industry will play its part in advancing world friendship."[77] The industry implied that as American-made films became increasingly global texts and commodities, domestic and international audiences alike would become more world-minded and tolerant citizens while remaining reassuringly under the cultural influence of the United States.

While educators were less sanguine in their assessment of the educational value of Hollywood films, many echoed the industry's utopian bromides when discussing the social and educational possibilities of the medium in general.[78] Though continually critical of the industry's slowness in providing titles that "correlated to the curriculum," educators frequently referred to the global, universal, and experiential nature of cinema as one of its essential and redeeming virtues. In writings, film reviews, and speeches, they claimed that films could "bring the world" to provincial Americans, and serve in at least some instructive capacity as a "universal language," "graphic Esperanto," source of "vicarious experience," agent of "international sympathy," and "magic carpet" that could take students "globe-trotting for a dime."[79] According to McAteer, who by then had led visual education programs in the Los Angeles City schools and at the University of California at Los Angeles, the experiential and worldly quality of movies made them useful to educators even when their content was exaggerated, incorrect, or otherwise "educationally valueless":

> Countless amounts of time, trouble and money have been expended by producers to film the Arctic and Sahara wastes, the pinnacles of the Alps, the depths of tropical jungles, the palaces of royalty, and the huts of distant and unknown peoples. It may be granted that, in many cases, this has been done merely to add realism to some fanciful and educationally valueless scenario. Nevertheless, it has provided for the viewers vicarious experiences of travel and association with many peoples of the world. Thus we can see that motion pictures can bring to the classroom objects and scenes which would be impossible of direct examination by the student body.[80]

Indeed, educators often claimed that films could serve as not only a substitute for but also a safer alternative to foreign travels and encounters.

Films could effectively deliver to students the useful parts of international experience while filtering out any undesirable aspects of face-to-face contact with racial and ethnic others. According to one educator in 1917, when compared to actual physical travel, film was "much cheaper and free from risk of life, discomfort of odors, atmospheric conditions, hardship and unpleasant experiences with strange people, dangers from aerial, submarine and other means of transportation, poisonous insects, wild beasts, and uncivilized man."[81] These sentiments were echoed in 1922 by Ruth Whitfield, a teacher in the renowned progressive school system of Winnetka, Illinois. Proposing "screen travels" as an innovative method for teaching international understanding in the schools, Whitfield wrote that films and lantern slides could provide the intellectual benefits of international journeys without the "dangers and discomforts." Citing the travel films and illustrated lectures produced by prominent white, male lecturers such as Burton Holmes and Edward M. Newman, Whitfield suggested that students could learn all they needed to about foreign cultures by following the edited journeys of experts. "When following the picture trail of a travel lecturer or news photographer, the ordinary currents of one's life are not interfered with, one is safe and comfortable and provided only with what is truly interesting and significant."[82]

Movies became somewhat more common in American schools during the 1930s. This was due to a number of city school and university extension systems investing in motion picture equipment, the growth of the Department of Visual Instruction (DVI) and its de facto journal, *Educational Screen*, and the establishment of teacher training courses in visual education at normal schools and universities.[83] A spate of new educational film companies, including Fox Films Corporation, Eastman Teaching Films (an extension of the Eastman Kodak Company), Bray Pictures Corporation, and Ideal Pictures Corporation, produced films on a variety of academic subjects. But as Paul Saettler notes, many of these companies struggled, halted production, or folded before World War II due to the financial constraints of the Great Depression and insufficient demand.[84] At the same time, new collaborations between educational groups and the theatrical film industry, producing the *Secrets of Success* (1934–1936) and *Human Relations* series (1935–1942), supplied some high schools with edited excerpts of theatrical films to promote group discussion of civics, social issues, and character education.[85] When the famous Payne Fund studies, conducted between

1929 and 1932, led by researchers at Ohio State University, and published between 1933 and 1935, found that commercial motion pictures directly influenced the attitudes and behaviors of youths, the industry quickly responded by adopting the Motion Picture Production Code—commonly referred to as the Hays Code—a set of self-regulatory guidelines to ensure moral decency in films. Together, these maneuvers by the film industry helped to placate educators and weaken the censorship movement.

Nevertheless, by the late 1930s the rates of film adoption in schools remained far below what educational and industry advocates hoped for. Throughout the decade, advocates of visual education lamented the embarrassing "lag" between the rapid development of theatrical motion picture technology for popular consumption and the haphazard, outdated use of the technology in schools.[86] The slow and uneven uptake of film in schools stemmed from a combination of factors, including cautious reluctance among school boards and administrators, the financial constraints of the Great Depression, an inadequate supply of scholastic films, and rapid technological changes, such as the introduction of sound film, that rendered some equipment obsolete.[87] The lackluster use was confirmed by a 1936 national survey of nine thousand school systems, representing some eighty-two thousand schools, conducted by Cline Koon, an audiovisual specialist, for the US Office of Education. The study showed that fewer than half of the nation's schools were equipped with sufficient electricity to show films. The responding school systems owned only 9,304 motion picture projectors between them, of which only 793 were equipped for sound. Schools owned and used fewer film projectors than radio sets (11,500) or lantern slide projectors (17,040). "It would appear from this that while mechanical improvements in the commercial entertainment field march steadily forward," lamented Koon, "the use of these [motion picture] improvements, so necessary for better transmission of modern educational ideas in a modern way, lags far behind in our nation's schools."[88]

For Koon, this "lag" in the educational uptake of film had implications not only for daily instruction but also for the survival of democracy and liberal citizenship in a time of rising fascism in Europe. Koon had recently authored another major report on behalf of the Office of Education—in conjunction with the National Congress of Parents and Teachers, the Pan-American Union, and Erpi Picture Consultants—that devoted an entire chapter to the importance of motion pictures in promoting international understanding. In that 1934 report, Koon optimistically hailed motion pictures as "the universal educators, the Esperantists of the universe. Upon motion pictures, more than on diplomats, depend the people of the world for their information concerning the lives, traits of character, and public policies of people in other countries. The cinema has important interracial implications and responsibilities. World-peace must be based on world acquaintance."[89] But with the disappointing findings of his 1936 survey, Koon warned that the schools' lag behind the general culture in adopting this new technology had serious implications for their duty to prepare students for modern citizenship in an interconnected and turbulent world. "Present-day social problems arising out of the complexities of our times will be effectively solved only if the teacher is given every aid created by the mind of man to combat them."[90]

Radio

The central and dominant aim of education by radio is to bring the world to the classroom, to make universally available the services of the finest teachers, the inspiration of the greatest leaders and the educative power of unfolding world events which through the radio may come as a vibrant and challenging textbook of the air.

—Benjamin Darrow[91]

Promotional discourses about virtual travel and world-minded citizenship were leveraged to make room for not only visual technologies but also radio in schools. Before the Communications Act of 1934, which allocated the majority of the radio spectrum to commercial broadcasters, the airwaves were crowded with the sounds of a heterogeneous mix of amateur, independent, and commercial broadcasters, including universities, newspapers, religious and educational groups, and department stores.[92] Despite their varied agendas, many of these early broadcasters produced educational programs in an effort to publicize their brands and impress on the public the social value of radio. Radio was described as creating "pictures in the mind," making it subject to many of the same lofty discourses about visual education and virtual experience as visual aids. Unlike lantern slides or motion pictures, however, proponents touted radio as a tool for mass instruction and distance education that could educate large and dispersed groups of learners simultaneously, whether within a single school building or entire school system, or across a region. Many regarded radio as a cost-effective medium of instruction that could strengthen the nation by transmitting the expert voices of "master teachers" to mass audiences, reducing educational disparities between urban and rural communities.[93]

To explore the potentials of educational broadcasting, several "schools of the air" were formed in the 1920s and 1930s with the backing of universities, corporations, philanthropic foundations, and school districts. Creators of schools of the air envisioned these programs as benefiting not only schoolchildren but the entire populace too, particularly immigrant families and rural audiences that they regarded as isolated from urban centers of education and uplift.[94] In 1924, the powerful retailer Sears, Roebuck and Co. sponsored the *Little Red Schoolhouse of the Air* on its fledgling radio station, WLS, in Chicago. The *Little Red Schoolhouse of the Air* transmitted lessons in geography, art and music appreciation, and science to an estimated twenty-eight thousand students across the Midwest. The original director, or "schoolmaster," of the program, Benjamin "Uncle Ben" Darrow, became a vocal proponent of radio instruction as an instrument of citizenship education and social uplift, claiming that by transmitting experiences of government, travel, and history to the masses, radio could serve as "an agency—a magic link—of the world and the classroom." Darrow's dream was to establish a federally funded "National School of the Air." While his aspiration was dashed with the legalization of a commercially dominated broadcasting system in the 1930s, he was successful in galvanizing a school of the air movement at the state level. Between 1927 and 1937, he directed the Ohio School of the Air, which was based at Ohio State University, and operated with financial support from the Payne Fund, Ohio Department of Education, and US Office of Education. The Ohio School of the Air broadcast lessons in civics, physical education, literature, and geography on weekday afternoons, reaching seventy-five thousand students in its first year. Though it suspended activity in 1937 due to declining support for state-funded radio, the school inspired several imitators in other states, including the Wisconsin School of the Air and Portland School of the Air, which continued broadcasting for decades.[95]

Similar to early educational film, one of the most prominent genres of early educational radio programming was the radio travelogue or "travel talk." With low production costs—requiring only an imaginative script, engaging lecturer, and ideally some music or sound effects—and high levels of interest reported from students and general audiences alike, educational and commercial broadcasters produced scores of talks on US and world geography in the 1920s and 1930s. Travel talks typically relied on a combination of informational lecture and dramatic narrative to take listeners on "imaginary journeys" through foreign lands. San Francisco's KGO station, established by General Electric in 1923 to promote the sale of radio sets, pioneered the genre with its "Geography by Radio" lessons in 1925. In these radio lessons, a male narrator called "Old Man of the Rivers" took a boy and girl on a sonic tour of the rivers of the world through storytelling and regional music.[96] Two years later, station WMAQ in Chicago, owned by the *Chicago Daily News* newspaper, drew inspiration from popular lantern slide travelogues to produce an ambitious kind of AV broadcast, later termed "illustrated radio," for the city's public schools. Using sets of colored lantern slides provided by the *Daily News*, teachers across the city projected identical slide shows of travel imagery in their classrooms while students listened to a geography lesson broadcast over the airwaves by a popular children's adventure author.[97] A significant development in both multimedia instruction and AV broadcasting before television, the program represented, as one educator reported in the *Educational Screen*, a promising breakthrough in "eye and ear instruction." Her account testified to a widely held belief in educational circles that radio, like cinema, could "bring the world" to students as well as "democratize" global experience through visualization, sensory appeal, and technological modernity. "What has been true for the eye since the perfection of the camera, the democratization of the awe-inspiring and beautiful sights of the earth, is now becoming true for the ear, . . . and this in the schoolroom."[98]

After commercial broadcasters gained control of the spectrum in 1934, educational programs about travel remained on the air as the networks sought to prove to the public that they could be trusted with stewardship of the airwaves. Educational radio advocates, their hopes for a national system of educational radio fading, turned their attention toward generating greater public demand for high-quality educational programming over the networks. Darrow, in a radio address sponsored by the National Congress of Parents and Teachers and broadcast over NBC in 1935, urged parents and teachers to "control the radio dial" and demand programming that not just entertained but also instructed children. Drawing from surveys he had conducted for the Ohio School of the Air, Darrow concluded that "travelogs," along with dramatizations of popular books and historical figures, were the programs best suited to this objective and happened to be what children most wanted to hear. "Whenever you see your son or daughter with eyes cast a thousand miles into the distance, the son, and perhaps even the daughter, is daydreaming of an airplane trip across the heart of Africa, or across the Russian steppes. They want more travelogs," he said.[99]

Because travelogues could easily weave together entertaining and educational content with messages about citizenship as well as national and global affairs, they offered a unique space on the airwaves in which various educational, commercial, and governmental interests could collaborate to produce programming that advanced their respective goals. In 1936, the Ohio School of the Air produced a series of travelogues through a partnership with the Federal Radio Workshop under the Radio Education Project, an initiative of the Roosevelt administration to allocate emergency relief funds from the Works Progress Administration (WPA) to develop limited educational programming on commercial stations. The Federal Radio Workshop employed a group of young educators in Cincinnati to write and perform a series of travelogues for schoolchildren on the history, geography, and industries of Latin America. This "travelog" program directly aligned with the objectives of Franklin Delano Roosevelt's Good Neighbor Policy, which sought to replace decades of US military interventionism in Latin America with a "softer" strategy of promoting cultural exchange and trade between the regions. True to these themes, the Ohio School of the Air travelogues mixed lessons on Latin American history and culture with paternalistic messages about Americans' responsibility to stimulate the economic development of their neighbors to the south. As a teacher's guide for the program explained, "There is so much that the well populated, strong and vigorous United States could do to aid in the development of these sister republics that every one of us needs to have our attention riveted upon them." A list of discussion questions further encouraged students to think of Latin America in terms of underdevelopment and unrealized economic potential: "What is taking place in Mexico that gives us hope that within another generation Mexico may have advanced a long way toward a better income and a higher standard of living?" Roosevelt's policy objectives were so salient in Federal Radio Workshop programming that the radio industry's trade journal *Broadcasting* complained it was allowing "political themes to creep in" to the commercial radio landscape.[100]

In the same year that the Latin American travelogues aired on the Ohio School of the Air, NBC created its own travelogue-inspired program called *The World Is Yours*, produced in conjunction with the Smithsonian Institution and US Office of Education, also under the Radio Education Project. Airing between 1936 and 1942 on weekend afternoons, *The World Is Yours* was the most popular and longest-running of several experimental programs produced by the radio networks, in partnership with the Office of Education, to develop educational content for commercial radio.[101] Along with *Americans All, Immigrants All* (CBS, 1938–1939), a series celebrating the contributions of immigrants to American history and culture, *The World Is Yours* was touted by the networks, educators, and government officials alike as evidence that dramatized educational programming could hold the public's interest while instructing people on issues of history, civics, democracy, and international affairs.[102]

Narrated by a grandfatherly explorer character called "Old Timer," *The World Is Yours* explored a mix of topics in natural history, science, geography, and the arts. Episodes were given adventurous titles, such as "The Lincoln Legend," "De Soto's Search for El Dorado," "Head Hunters," and "The Inca Empire of the Sun." They were enlivened with dramatic dialogue and music, directed by Rudolf Schramm, who had scored the popular documentary film *Nanook of the North*.[103] Promotional brochures for the program depicted a pair of hands holding the entire globe, a suitcase covered in stickers from destinations around the world, and the heavy doors of the Smithsonian swinging open, promising access to the museum's "treasure rooms" and urging audiences to "make the world more completely YOURS!" These materials aligned closely with the promotional discourse, touted by educational radio advocates like Darrow for over a decade, that radio could "pick the lock" on elite institutions, such as universities and museums, and bring worldly knowledge to the masses.[104]

But the program's reliance on "ultra-popularization" techniques such as dramatic narrative, music, and fictional characters concerned some Smithsonian curators, who worried that it trivialized science and history, and contributed to simplistic stereotyping of complex civilizations.[105] An analysis of the scripts suggest that these concerns were not unfounded. An episode on "Primitive Music" (May 8, 1938), for example, opens with a pair of American teenagers dancing to swing music at a party. There they

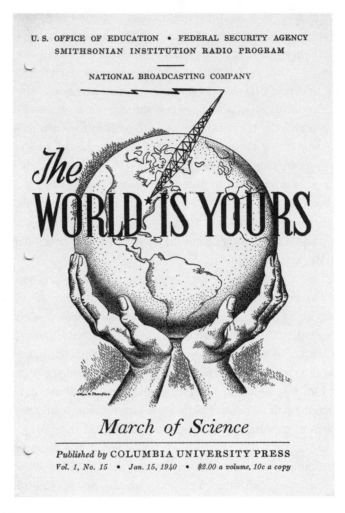

Figure 2.7
Cover of the supplementary material for *The World Is Yours*, issued January 15, 1940, Record Unit 83, Editorial and Publications Division, Records, 1847–1966, SIA 2007-019, Smithsonian Institution.

encounter Old Timer, recently returned from a trip around the world, and sporting a "tanned face" and linens. He remarks that the music at the party sounds like "something out of Africa" and "reminds me of Zulu ceremonials I've seen." Old Timer proceeds to trace the origins of music and rhythm back to "before the dawn of history," in a time of "naked savages" when "ancient jungles echoed to the sound of booming drums." The episode's narrative reinforced racist understandings of African music as "primitive," uncivilized, and frozen in premodernity, while linking the contemporary genre of swing—which was pioneered by African American performers—to this fiction.[106]

Drawing on the travelogue tradition, episodes of *The World Is Yours* frequently featured an experienced narrator taking another character on a "tour" of a foreign place, offering explanations about its history, culture, and customs. In "Eskimos" (undated), the voice of Old Timer was replaced by that of another white male explorer figure, a missionary named Father Patrick, who explains Inuit culture to an American airplane mechanic, newly stationed in Alaska, named Tim. In an episode that repeatedly describes the Eskimos as "the loneliest people in the world," Father Patrick is introduced as a "loneliness fighter who every day flies over the Alaskan tundra land bringing cheer" to them. When the men receive an emergency call over the radio that a local man has been injured by a walrus, Father Patrick and Tim fly to the rescue. On landing, Tim is surprised to see that the wife of the injured man, Minooko, is stoic as they prepare to airlift her husband to a hospital. He interprets her stoicism as a sign of ungratefulness. "You misunderstand, Tim. She is overjoyed!" explains Father Patrick. He transitions into a history lesson about the Eskimos' bleak and lonely existence. Their "struggle" was worsened when "the white man came and added their burden," referring to the Russians who "weakened their ranks with fiery liquors they had never known before," and "took advantage of their childish trust and skill as hunters and trappers and robbed and cheated from them what they caught." When the United States bought Alaska from the Russians, he explained further, "Profiteering started again . . . murder and the stream of liquor, but before long Uncle Sam stepped in and stopped all that. Schools were started, financial aid and hospitals were given them, and most important of all, something the Eskimos had never known . . . kindness. These lonely, isolated people—forgotten, ignored for centuries— were knowing kindness." The Eskimo woman's stoicism was not a sign of ungratefulness, Father Patrick remarked, but rather a sign of weary gratitude, like that of "a little beaten dog who licks his protector's hand." "See, now, what she meant, Tim?" he went on. "She was thankful that anyone even thought of helping her man. . . . That we even thought of bringing cheer to her lonely existence is enough for her."[107] Through narratives like these, *The World Is Yours* attempted to educate the public on world cultures, but did so in a manner that thoroughly exoticized, infantilized, and essentialized indigenous and non-Western experiences through the perspective of white male explorers. The program routinely reinforced racist and paternalistic assumptions that the non-Western world needed the United States' help to develop toward a more civilized state.

While the program's tendency to dramatize and distill entire cultures bothered some curators, for audiences, this mode of presenting natural history was a hit. Within months of airing, the Smithsonian received over three thousand pieces of fan mail, many coming from "school teachers and others engaged in educational work."[108] In a letter addressed to Old Timer, one fan wrote that listening to *The World Is Yours* was "the first time I've ever been educated and entertained over the radio at one time." In another letter, a high school teacher commented, "If more educational programs bothered to concoct a script which was painless to listen to, more of us would bother to listen."[109] Though the program aired on weekends, a number of teachers incorporated it into their lessons. Some assigned students to listen to the broadcasts as homework, while others had students re-create the broadcasts in school using scripts provided by the Office of Education's newly formed Educational Radio Script Exchange.

Despite the popularity of the program, production ceased in 1942 due to the war. Congress cut off funding for the Radio Education Project in 1940, effectively ending the US Office of Education's activity in radio production. As Saettler writes, the takeaway lesson from *The World Is Yours* and other public service programs from the late 1930s was "that radio had greater possibilities for the stimulation of learning than it had for systematic teaching."[110] While educators' dream of a robust and federally supported system of educational radio never materialized, radio programs about world travel and citizenship, a key genre in the development of early educational broadcasting in the 1920s, remained a fixture of the airwaves during the early network era. Demonstrating that radio could simultaneously entertain audiences while educating them about the world and the United States' increasingly powerful role within it, programs about travel and cross-cultural encounter, however reductive in their presentation of diversity, proved to be an influential transitional genre that helped legitimate the medium's place in American society and burnish the public image of the commercial broadcasting system.

Emerging Media and World-Minded Americans

In the first four decades of the twentieth century, media producers and advocates of visual education drew on techno-utopian and global discourses to make a case for bringing new technology into the classroom. A recurring argument, made across time and technologies, was that audio and visual devices could expose students to the complex world beyond the classroom, and mold them into a more world-minded yet ideologically unified American citizenry. Channeling long-standing understandings of visual media as a substitute for travel alongside emerging, progressivist ideals about the primacy of experience in learning and the utility of technology in solving social problems, promoters endowed new media technologies with the capacity to efficiently "bring the world" to young people. In this view, technology could vividly, economically, and safely simulate firsthand encounters with geographic as well as cultural diversity to better prepare young people for informed citizenship in an increasingly diverse, industrialized, and interconnected society.

Ironically, it was the United States' entry into the violent global conflict of the Second World War that provided both the jolt to the fledgling educational technology industry and political justification for making educational technology mainstream. The urgency of counteracting Axis propaganda, and training large numbers of troops and civilians for the war effort, led the government, armed forces, and educational groups to aggressively produce and disseminate media technologies, including film, radio, slides, and filmstrips, for educational purposes. The government enlisted many in the professional educational technology sector, including commercial media producers and educational researchers, to produce instructional aids and propaganda for the war effort. According to Saettler, during the war the armed forces made six times as many films as had ever been made before for instructional purposes.[111] Schools could secure these films at no cost, in addition to films produced by other federal agencies, such as the Department of Agriculture, Office of War Information, and Office of the Coordinator of Inter-American Affairs. The war also hastened the material development of audio and visual technology, making devices cheaper, more effective, more mobile, and easier to operate. Together, these changes paved the way for a postwar era in which AV equipment and techniques became commonplace in schools.[112]

After the war, hundreds of technologists who were trained in wartime joined the NEA's Department of Visual Instruction, causing its membership to nearly double to over a thousand members by 1946. Within a decade, membership tripled to three thousand.[113] What had formerly been known as the visual education movement quickly gave way to the professionalized field of AV education—a transition made official when the DVI changed its name to the Department of Audio-Visual Instruction (DAVI) in 1947. Armed with a growing body of empirical research on AV methods in teaching and a pragmatic, "systems"-oriented approach to incorporating technologies into the curriculum, the AV profession was less inclined toward the sweeping rhetoric of social uplift used by interwar media producers and visual education advocates to justify technology in education.[114]

Nevertheless, a discursive theme that continued to appear in promotions for AV educational technology, during and after the war, was the utility of technology in exposing young people to the international world, and "readying" them for "the responsibilities of citizenship" in an even more complex and interconnected future.[115] As international understanding and intercultural education became increasingly salient priorities for American educators during the war, AV producers and professionals alike underscored the unique ability of audio and visual aids to make students more world-minded yet patriotic citizens. As William Lewin, a New Jersey teacher and chair of the NEA's Committee on Motion Pictures for the Department of Secondary Teachers, wrote in the *Educational Screen* in 1943, movies and other media must play an important role in "the great task of democratizing understanding" after the war. He argued that "one of the chief tests of success for administrations of visual instruction must, therefore, increasingly be a measure of their ability to provide for the utilization of audiovisual materials, whether in school or out, for the development of enduring peace and prosperity through the cultivation of world citizenship, side by side with local pride and justifiable patriotism."[116]

Consistent in the promotional rhetoric for all manner of instructional media was a dual emphasis on its ability to unify and uplift Americans around a common set of civic values while turning their attention to the changing world beyond their shores. Educators were encouraged to harness technology to help their students learn more about the world while taking stock of how much this technology was making the world more like them. The incorporation of educational technology into US schools was thus deeply bound up with social and political anxieties about international interdependence, demographic diversity, and intercultural contact as well as conflict. Educators' and media makers' discourses revealed ambiguous beliefs about the progressive and peace-building power of expanding the representation of racial and ethnic diversity in schools, even as these representations reinforced reductive, Anglo-Amerocentric definitions of citizenship and national power. Early twentieth-century promoters of visual education thus set in motion the assumption—still present in today's discussions of global and mediated learning—that teaching with technology could not only make Americans more cosmopolitan citizens of the world but also stronger citizens of the nation and defenders of its interests.

3   Coming to Our Senses: Cosmopolitan Technologies for Cosmopolitan Citizenship

The other day I noticed four or five sixth-grade boys who were staying after school to put away lantern, slides, and chairs after a visual education period. They didn't start for home immediately. They clustered around a table where there were a lot of sea curios. They held the great shells to their ears. They rubbed the rough coral against their cheeks. One even touched it with his tongue. . . . They held the seaweed to the light, they pricked their fingers with the sea-horse and the sea-urchins. Shall I confess the appalling truth? Not one boy turned to the map to locate the spot where these treasures were found. Not one came to me with a flood of eager questions. They were absorbed in the immediate sensory experience. They were very still, very intent, and there were, if I am not mistaken, dreams in their eyes. Poor youngsters, fed upon words when they are hungry for concrete realities.

—Edna Collamore, "The Why of Geography Exhibits," 1928

In the early days of the visual education movement, educators often echoed the worldly and utopian claims of commercial media producers when discussing the instructive potentials of new media and sensory aids. Though they challenged the academic merit of some slides, stereographs, films, and other ready-made technologies, few questioned the general assumption that such devices could reliably "bring the world" to children and serve as efficient substitutes for experience. But after World War I, changes in the communication landscape and pedagogical theory led a number of prominent, media-minded educators to diverge from these promotional scripts. Among them were leaders of the newly formed professional organizations for visual educators, including Anna Verona Dorris, president of the NEA's DVI (1927–1929), F. Dean McClusky, president of the National Academy of Visual Instruction (1930–1931), Charles Hoban Sr., DVI president

(1932–1933), and Edgar Dale, another president of the DVI (1937–1938).[1] While enthusiastic about the potentials of teaching with new technology, this interwar cohort of advocates and researchers expressed dissatisfaction with the assumption that it could automatically transmit experience to students, noting its connections to commercialism, passive consumerism, and "mindless" entertainment. Instead of unquestioning reliance on technology, they urged their fellow educators to adopt more "active," varied, multisensory, and critical uses of media in schools. By the end of World War II, they helped to establish a number of techniques to realign visual education with the tenets of progressive pedagogical theory and the changing curriculum of the postwar age.

This chapter argues that through their advocacy for more participatory and varied engagements with media in school, interwar educators laid the foundation for what is now a familiar—and still widely sought-after—educational ideal in the twenty-first century: the cultivation of a modern, media-enhanced classroom in which students learn to act not as passive spectators but rather active critics of and engaged participants in their multimediated, globally connected society.[2] Alarmed by the rising social influence of mass media industries and the threat of propaganda, these educators called for schools to resist dependency on ready-made media products, and cultivate instead their own localized mixes of high- and low-tech aids, marking such activities as crucial to developing the senses, teaching world-minded citizenship, and safeguarding democracy. Media of all varieties were to be seen not as finished *products* to be passively imported into the classroom but rather raw educational "materials" to be acted on, modified, mixed, appraised, and even refashioned by discerning citizen-consumers. This philosophy, which reflected a combination of nascent anxiety about the spread of mechanized communication and a long history of Western educational reforms to promote sensory education, ushered a series of hands-on learning devices into the twentieth-century school, including school museums and exhibits, homegrown media collections, bulletin boards and scrapbooks, and lessons in media making and criticism. By prioritizing "active," "direct," and "concrete" approaches to mediated learning during the formation of the instructional technology movement, reformers articulated a new vision of young people's media reception as both a radically critical and thoroughly disciplined educational activity, and the classroom as not an "adjunct of the theater" but instead a utopian space of democratic communication, multisensory learning, and citizenship formation.[3]

These ideas were codified into mainstream American education after World War II with the publication in 1946 of *Audio-Visual Methods in Teaching*, a textbook for teachers written by Dale, a professor of curriculum at the Bureau of Educational Research at Ohio State University. Dale is primarily remembered by media scholars for his early career contributions to the Payne Fund studies, which explored the social and behavioral effects of motion pictures on youths, and his subsequent leadership of the film appreciation movement, a forerunner to film studies and media literacy education, in the 1930s. But as Charles Acland has noted, Dale's vision of media literacy was not limited to the medium of film, and reflected instead the increasingly multimodal and integrated nature of communication over the duration of his career from the early 1930s to the 1970s. The hallmark of Dale's book was the "cone of experience," a guide to educational aids that connected the progressive movement's long-standing commitments to direct, participatory experiences—the kind obtained through building a model, going on a field trip, or performing in a pageant—with the more symbolic, abstract, and virtual experiences offered by the era's emerging media. The cone of experience was the culmination of Dale's ruminations in the late 1930s and early 1940s that schools should prepare students to engage with not only the ascendant mass media of the day—film, radio, and newspapers—but also humbler sources of everyday information, including exhibits, objects, speeches, dramatizations, and other homely materials. He was, as Acland writes, "a media literacy advocate before there was such a term" and approached "discrimination in all media use as key to a modern education."[4]

This chapter begins by tracing the long-developing preoccupations with sensory learning that took shape within the progressive and visual education movements, and ultimately shaped Dale's writings. These concerns prompted a number of homegrown, recombinant, and intermedial practices to enter American classrooms after the turn of the century, such as the use of school museums, "realia" collections, bulletin boards, scrapbooks, and opaque projection. Educators valorized these varied, localized, user-centric, and multisensory uses of media as a way to make children literate in new informational practices while "inoculating" them against the deindividuating effects of industrialization, mass communication, and propaganda.

Against a backdrop of proliferating mass communications, mechanized warfare, and concerns that uncritical consumption of media could bring about an undemocratic "mob mind," Dale and his contemporaries began to speak of cosmopolitan collections of media and materials in the classroom as a key to forming cosmopolitan subjectivities. As his research unfolded in the 1930s, Dale's pedagogical interests in how children read as well as obtain meaning from their environment began to intertwine with a growing social concern about the preponderance of degrading racial and ethnic stereotypes in popular media in American culture and the rise of antidemocratic mass media in Nazi Germany. Concerned that Americans' "uncritical attitude" toward media made the country vulnerable to the appeals of fascists and propagandists, he began to yoke his vision of a more participatory AV instruction to liberal ideals of international understanding, antiprejudice, and free speech. By the postwar period, when he took on roles as an adviser on communication and technology for the United Nations Educational, Scientific and Cultural Organization (UNESCO), and became an internationally recognized figure in AV education, Dale had helped make mainstream the belief that the broad and mindful use of multiple forms of media technology in school was not only good for instruction but good for global democracy too.[5]

Ironically, the efforts by Dale and his contemporaries to foster more critical and child-driven applications of media created the ideal conditions for more technological products to enter schools in the latter half of the century. While reimagining "active" AV instruction as a democratic alternative and counterweight to the culture of passive consumerism and mass persuasion that burgeoned beyond the school walls, Dale and his cohort built a pedagogical rationale for embracing the multimediated, participatory, and interactive logic of the information age that was just over the horizon.[6] Though they cautioned against thoughtlessly importing technology into instruction, they nevertheless advanced a different kind of technological idealism that asserted that empowering young people through frequent and varied engagements with technology, rather than restricting media use or regulating industries, was the key to generating a healthier and more democratic communication landscape. Additionally, while Dale was one of the first prominent educational theorists to argue that teaching media criticism in schools could improve racial and intercultural relations in the United States and around the world, his recommendations were limited to helping white students identify and reject simplistic racial and ethnic stereotypes in film, radio, and the press. Optimistic that Americans' democratic spirit would, with proper education, lead them to make principled media choices at the box office and newsstand, and on the radio dial, Dale's vision of social reform through media literacy education avoided confronting the underlying structural inequalities, internalized biases, and policies that produced these problematic representations in the first place.

Engaging Children's Senses in an Age of Machines

The end of World War I launched what Paul Saettler calls the "decade of growth" (1918–1928) in the visual education movement. The mass mobilization of visual aids in the war effort, including the use of posters, slides, and films for military and propaganda purposes, convinced many in the educational community of new technology's instructional value. These years saw a surge of efforts to organize, expand, and discipline the use of these devices in education. In 1923, the NEA established the DVI, a group of public school teachers and administrators interested in exploring new methods of visual education. The *Educational Screen*, one of several journals dedicated to visual education that formed shortly after the war, became its unofficial publication.[7] In the 1930s, the administrative leadership of the DVI shifted from school professionals to university researchers and educators, reflecting the growing interest in educational technology within higher education. In 1932, the DVI absorbed two similar organizations, the National Academy of Visual Instruction and Visual Instruction Association of America, and increased its membership to some six hundred members over the next decade. While the market for educational technology struggled throughout this period, particularly during the Great Depression, few American educators escaped encounters with new instructional devices—whether films, slides, opaque projectors, filmstrips, stereoscopes, record players, or radio—in their schools, educational literature, or the popular press. A central question that motivated the nascent visual education movement was not only how to promote the greater use of these devices in schools but also how to promote more "intelligent and systematic use" of them in a way that would distinguish them from the world of popular cultural amusements—while still harnessing some of their interest and appeal.[8]

This vision had begun gaining traction during the war at the NEA's annual meetings, where the Committee on Visual Instruction—a predecessor to the DVI—convened under the organization's Department of Science Instruction in 1916. While the educational possibilities of photography and film had been a topic of discussion at the NEA for years, the formation of a committee gave leaders a venue in which to link their interest in these popular technologies to demands that they be applied to "serious educational ends."[9] The meeting was held in New York City at the American Museum of Natural History, whose recent launch of a citywide school service loaning lantern slides, boxed exhibits, and other visual materials to public schools had made it a revered trailblazer among media-minded educators. Outlining a vision for the new committee, Edward Stitt, the district superintendent of the New York City schools, argued that the time had come to "enlarge the plan and scope" of visual education beyond the most talked-about medium of the day—motion pictures—and incorporate a broader range of materials into the classroom. In addition to adopting mechanical aids like films, slides, and stereographs, he contended, educators should make liberal use of nonmechanical devices such as maps, charts, models, blackboards, illustrations, postcards, magazines, newspapers, exhibits, school museum collections, clay modeling, sand tables, homemade models, and visits to museums. By championing this heterogeneous mix of sensory aids and activities, Stitt urged educators to prevent visual education from becoming "passive" and bring it into line with the central tenet of progressive pedagogy: "active" learning. "The next step in progressive pedagogic development," he declared, "will be a release from the passive reception of the wonders of film reproductions, by enlisting the active energies of the pupils so as to awaken their self-activity."[10]

The early calls to control and contain the encroachment of "passive" media into the classroom stemmed, in part, from nascent anxieties about the rise of mass media and machine age communication. Of particular concern was the skyrocketing popularity of motion pictures, which according to Charles Crumly, an Alabama science teacher, produced "evil effects" in children, including an inability to concentrate and "nervousness often bordering upon hysteria." School officials and university researchers also fretted about how technology would impact the teaching profession. As one educator warned in a 1920 essay titled "Visual Education—Spur or Sedative?" the schoolteacher who relied too much on teaching with newfangled devices, like films and lantern slides, risked becoming "a machine-driven puppet instead of a master."[11]

But these anxieties were not wholly new, stemming from a long history of Western efforts to reform education through the engagement of the senses. Indeed, many educators, including Superintendent Stitt, referred to this centuries-old history of sensory pedagogy to suggest that the growing popularity of visual instruction in the new century was the culmination of generations of hard-won educational progress rather than a fad.[12] Visual educators commonly cited as the movement's patron saint the Czech educator John Amos Comenius (Jan Amos Komenský, 1592–1670), who in 1658 published what is often called the first children's picture book, *Orbis Sensualium Pictus* (variously translated as "The Visible World" and "A World of Things Obvious to the Senses, Drawn in Pictures"). An encyclopedic compendium of woodcut images pertaining to geography, zoology, botany, theology, and everyday life, *Orbis* reflected Comenius's educational philosophy that humankind's relationship to the world could be studied more effectively through the senses, especially vision, than via the written word alone.[13] Over a century later, the Swiss Romantic pedagogue Johann Heinrich Pestalozzi (1746–1827) further advanced the concept of sensory learning, described as *Anschauung*, or "sense training," in European education. Known for his motto of "learning by head, hand, and heart," Pestalozzi believed that children learned best through direct observation and firsthand experiences rather than through abstracted approaches like rote memorization or recitation.[14]

These developments paved the way for several educational movements in Europe and the United States that emphasized learning through the senses, and particularly observation. Against a backdrop of rapid industrialization, the object lesson, nature study, and picture study movements all emerged in the late nineteenth century in opposition to the mass educational techniques of rote, "mechanical learning" and verbalism. Object lessons stressed the close observation of material things—objects, natural specimens, and images—to illustrate abstract concepts and ideas. Relatedly, the nature study movement sought to compensate for urban students' lack of experience in the natural world through firsthand study of specimens and field excursions. Finally, an approach similar to visual education—picture study—gained many followers from the 1890s through the 1930s. Spurred by the growing availability of mass-reproduced images, picture

study involved the use of printed illustrations and reproductions of paintings to stimulate students' appreciation of art, and inspire creative activities
in other subjects such as geography, history, language, and reading.[15] As
Sarah Carter writes of object lessons, these sensory devices gained popularity in education due to their perceived ability to help students "bridge the
gap between the familiar and the foreign," using local, "concrete" examples
to enlarge the child's understanding of the abstract and complex world
beyond the classroom.[16]

Since many interwar visual educators saw themselves as inheritors of
this progressive educational tradition, the prospect of bringing new, mass-
produced media devices into schools presented both an opportunity and
threat. On the one hand, technologies like motion pictures, slides, and
radio sets offered ways to make the abstract concrete and efficiently bring
"experiences" of the wider world into the classroom. On the other hand,
their associations with passivity, commercialism, and mass entertainment
seemed out of line with the progressive doctrine about the importance of
active, child-centered learning and informed citizenship. For some, disciplining these technologies for educational use was a matter of institutional
and professional self-preservation. Since the turn of the century, newspapers had sensationalized claims that newfangled media technologies—from
stereoscopes to phonographs to motion pictures—were going to "revolutionize" education and transform the teaching profession, perhaps rendering teachers obsolete. A notable progenitor of this discourse was Thomas
Edison, discussed in the previous chapter, who made headlines in the 1910s
and 1920s by claiming that motion pictures would one day transform
schooling as well as shoulder much of the work of teachers.

While newspapers commonly painted teachers as provincial techno-
phobes resistant to change, many educators responded to the techno-
euphoric predictions by calling not for new devices to be banned from
the school altogether but rather for them to be disciplined according to
accepted pedagogical principles.[17] They began by making a distinction
between "passive" and "active" uses of visual aids, and frequently classified the use of motion pictures—the most talked-about technology of the
day—as an example of the former. As Crumly cautioned in 1919, "A child
engaged in merely observing the film is not active in the right way. While
no thoughtful person today would favor excluding the screen from the
schools, it must be realized that the use of it must be much more limited

than certain enthusiasts imagined." A similar sentiment was communicated the following year by US commissioner of education Philander P. Claxton. Addressing educators in the new journal *Visual Education*, he wrote, "Man is a thinking and speaking animal. Mere gazing at objects and pictures— even at the most interesting moving pictures—like a cow gazing a new gate, will not result in education, not even in knowledge. Knowledge and education are something more than mere sense impressions, and something very much more difficult to obtain."[18] For the next two decades, as visual educators set out to establish the most effective modes of using technology in the classroom, they echoed Claxton's and Crumly's calls. For many, teaching with media and technology would require adopting more diverse devices than just films, engaging more senses than merely vision, and perhaps most challenging of all, transforming students from passive consumers into active, civic-minded collectors, creators, and critics of media.

Championing Multiple Media, Challenging Industry Monoliths

The burgeoning educational interest in promoting a more participatory and multisensory instruction was rooted, in large part, in progressive educators' deep-seated unease with the encroachment of commercialism into schools and society. As the commercial film industry grew and began building partnerships with the educational community in the 1920s, some educators cheered the opportunity for greater input in movie production, while others chafed at the idea of a powerful industry exerting influence over education. Limited school budgets, furthermore, particularly during the Great Depression, slowed schools' efforts to incorporate motion pictures or other equipment into instruction. With film and radio skyrocketing in popularity among the general public, educators reasserted their professional roles and expertise by reminding each other, in pedagogical literature and at conferences, that visual education was defined not merely by newfangled machines but also by a vast array of other sensory devices and experiences to be orchestrated by the discerning teacher. As the *Educational Screen* editorialized in 1926,

> Visual instruction no more means films than it means actual objects, models, maps, charts, graphs, diagrams, posters, cartons, prints, cuts, photographs, stereographs, or slides. It means them all—properly adapted to and articulated with the particular teaching job in hand. Subjects best taught by motion should use films;

other subjects—and they are legion, should not. We do not use steam-rollers to crack nuts, nor teaspoons to dig canals.[19]

To some educators, the problem of privileging motion pictures and other popular visual technologies in the classroom was bound up with a growing cultural overreliance on vision. While united by a desire to challenge the primacy of verbalism in instruction, some feared that the pendulum would swing too far in the other direction and favor information delivery via the eye at the expense of the other senses. Joseph Weber, a professor of education at the University of Kansas, whose 1922 doctoral thesis at Columbia University was one of the first studies to demonstrate the effectiveness of visual aids in instruction, lambasted the visual education movement in 1928 for "placing the sense of vision on a pedestal and worshipping it as the golden calf." He urged practitioners to remember that "all the senses, more or less and in diverse combinations, co-operate in the steady accumulation of learning."[20] Some critics saw visual education as not a remedy for but instead a symptom of the rising hegemony of visual attractions in society. As journalist Harry Shaw griped in 1937, the growth of visual instruction in schools was only worsening a national crisis in which Americans were becoming increasingly "picture-minded" and less literary in their apprehension of knowledge. In his view, the technologies of cinema, pictorial magazines, and radio fostered a "passive receptivity" that "inevitably encourages mental laziness." Young people's voracious consumption of imagery—whether projected on a screen, printed on a magazine page, or conjured in the mind through compelling radio broadcasts—not only made them poorer pupils but also weaker citizens more vulnerable to the appeals of advertisers and propagandists.[21]

Addressing these concerns, a number of educational researchers in the late 1920s and 1930s called for a more pedagogically informed praxis of visual education in which teachers, students, and the curriculum would drive the use of devices, not the other way around. Importantly, these formulations made room for motion pictures in instruction, but stressed that any screenings in school should be anchored in a range of auxiliary activities designed to prioritize "active" experiences and critical thought. Dorris's *Visual Instruction in the Public Schools* (1928) was something of a transitional text in this regard, at various turns marveling at the totalizing "magic" of motion picture experiences while noting the need to "control and regulate visual education," particularly movies, to promote a "more pedagogical use of all visual aids." "If interest in pictures is confined to the mere information they convey," she warned, "their contribution is small indeed. Information should lead to thought and activity; activity should develop conscious strength, self-confidence, and increased command over the knowledge acquired."[22] These ideas became more formalized with the publication of *Visualizing the Curriculum* (1937), a popular textbook used in teacher education courses that was adapted from the dissertation research of Charles Hoban Jr., a motion picture consultant for the American Council on Education, along with his father, Charles Hoban Sr., the director of visual education for the state of Pennsylvania. Presenting a systematic and relational view of visual aids, the authors argued that each device's instructional value was determined not by its individual technical properties but rather its "integration" with other devices and "correlation" to an experience-based curriculum. "The heart does not beat apart from the expansion or contraction of the lungs, nor do the muscles flex apart from the bones of the skeleton," they wrote. "So, in the teaching process, instructional materials fit into a totality of experience, each contributing its particular element to the integrated whole." They proposed that materials should be mobilized according to a "scale of concreteness and abstraction," with the assumption—hearkening back to the ideas of Pestalozzi—that concrete learning should precede the abstract. In their view, the most concrete experiences came from school journeys and museum materials, while "graphic materials," such as maps, charts, and illustrations, were more abstract. Motion pictures and projected images, neither alpha nor omega, fell somewhere in the middle.[23]

This "concrete-abstract rationale" undergirded much of the emerging research on AV instruction in the 1930s, including Dale's work at Ohio State University that eventually led to his publication of *Audio-Visual Methods in Teaching* in 1946.[24] The book was organized around the concept that would come to define Dale's career: the cone of experience, a pyramid-shaped schema that recommended a balanced mix of audio and visual aids along with experiences in education, much as the food pyramid would later recommend foods. Similar to the Hobans, Dale's cone proposed moving "between the two extremes [of] direct experience and pure abstraction," but departed from earlier work by incorporating audio as well as visual aids into the process and prioritizing, at the foundation, the building up of children's "direct, purposeful experiences" in their immediate environment without any mediating technology. The cone was roughly laid out into three broad categories of experience, with those at the base emphasizing learning by *doing* (creating, working with models, and participating in plays, pageants, and tableaux), followed by *observing* (partaking in field trips, demonstrations, exhibits, motion pictures, radio, recordings, and still pictures), and finally *symbolizing* (interpreting visual and verbal symbols, such as maps, charts, graphs, and ultimately words). Calling for the full "integration" of these different sensory aids and experiences, the cone neatly crystallized the multisensory, experiential, and child-centered approaches to mediated learning that had been slowly building in the visual education literature since World War I. In essence, the cone stipulated that no single visual or audio aid was educationally superior or even sufficient in its own right; rather, they needed to be intentionally combined, controlled, and anchored in student-driven activity for their educational benefits to be fully realized.[25]

Dale worked through these ideas in the late 1930s in his *News Letter*, a monthly bulletin that he edited and distributed from the Bureau of Educational Research, with support from the Payne Fund, to thousands of educators in the United States and abroad for the duration of his career. Drawing frequently from John Dewey, Dale maintained that the purpose of schools was to provide a "rich base of experience" for pupils, particularly as the growth of mass media beyond the school walls meant that children were increasingly experiencing the world in a "vicarious, indirect" manner.[26] In a 1939 article called "Coming to Our Senses," Dale wrote that while learning at the "symbolic, mediated levels" was unavoidable in a modern society, "the symbols should always rest on a rich base of experience that the individual re-creates as intellectual and emotional responses to such symbols." He added,

> One basic danger confronts a world in which the indirect experiences of individuals are sharply increased by means of communication devices such as radio, movies, and the press. It is the hazard that as life is lived increasingly on the vicarious, indirect level, the symbol may be substituted for the experience of which it is merely a sign. . . . We may live our lives by proxy—seated in a comfortable chair, listening to, looking at, reading the experiences of others. . . . Teachers must not forget that children have eyes, ears, noses and muscles, and that they like to use

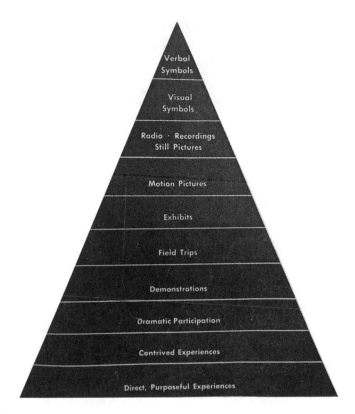

Figure 3.1
The cone of experience. Edgar Dale, *Audio-Visual Methods in Teaching* (New York: Dryden Press, 1946).

> them. . . . We must remember that the experiences which are participatory in nature, experiences for which we are responsible and for whose outcomes we share responsibility, are the ones which are truly educational.[27]

Rather than propose that children avoid or reduce their consumption of mediated information, the remedy that Dale eventually arrived at was that teachers should expose children to a multitude of media. Through carefully structured activities, children could build up their own base of direct, "full-bodied experience" in school, whether by visiting farms, creating art, making their own movies, or performing radio plays. Once established, these participatory experiences—"the bedrock of all education"—would enable them to more consciously and intelligently assimilate the streams of symbolic information that they encountered, in and out of school, through movies, radio, and other mass media forms.[28]

Further disciplining the practice of visual education and distinguishing it from the world of popular visual amusements, other interwar educators paired their warnings about the dangers of passive receptivity with calls for less vision-centric terminologies. "Visual education? Certainly," wrote Collamore in the *Journal of Geography* in 1928. "Also tactual, auditory, kinaesthetic, olfactory, and gustatory education. . . . Why neglect so many sources of approach to the brain? Keep ALL the sensory processes educationally active, then children will learn more easily and with less fatigue."[29] A growing chorus of educators called for retiring the term "visual aids" in favor of more inclusive alternatives, such as "perceptual aids" and "visual-sensory aids," before ultimately settling on "audio-visual aids" and "audio-visual education" in the mid-1940s. The shift from visual to audiovisual or AV education became official when the NEA's Department of Visual Instruction—then the largest organization of visual educators in the United States—changed its name to DAVI in 1947. In a parallel move, some visual educators argued for retiring the term "aids" altogether in favor of "materials" to foreground the user-driven activities that they regarded as necessary to activate any medium's instructive potential. As one teacher put it in 1945, whereas AV *aids* implied devices that merely summarized information for students, *materials* were "to be figuratively handled or manipulated, studied and learned for their own sake."[30] Through this semantic shift, educators hoped to precipitate a cultural one in which the educative power of media technology would reside not in the objects themselves but instead in the unique applications and experiences of the people who used them.

In their appeals to foster more varied and participatory uses of media in schools, visual educators often criticized the inadequacies of the ready-made media products that the burgeoning technology industry "foisted" on them. McClusky, a school administrator in New York State who authored several studies for the early visual education movement and served as president of the National Academy of Visual Instruction before its merger with the DVI in 1932, called for a new approach to visual education that would make students the primary producers of visual aids. In his view, children could learn more from scouring the "ten-cent store, the drug store, the waste heaps in any department store, printer's shop, carpenter's shop, or hardware store" for instructional materials than relying on media products peddled in trade catalogs. McClusky's vision for a more grassroots visual education corresponded to an upswell of anticorporate activity in American educational media and visual culture in the 1930s, as reformers lobbied to rein in the expanding radio broadcasting and film industries, educators encouraged each other to make do with limited resources during the Great Depression, and New Deal and philanthropic foundations spurred a wave of civic-minded media, murals, educational exhibits, and public art.[31]

McClusky found evidence of the pedagogical value of humble, child-made media at the 1928 Progressive Education Association conference, which was held New York City and featured Dewey as the keynote speaker. There, McClusky attended an exhibit at the Metropolitan Museum of Art that featured work by students from progressive schools across the state. Reporting in the *Educational Screen*, he marveled at the diversity of visual aids that students had created. With some aids made by hand, and others with the help of cameras and other mechanical devices, this "work of modern school children" epitomized the potentials of a visual education program centered on heterogeneous media and materials along with student activity:

> The exhibit hall was a sea of almost all types of visual materials. . . . One fourth grade sent its own "moving picture" of transportation on the Hudson; photographs of pageants, of apparatus, of dramatizations, and of activities were to be found; collections; scrapbooks; child-made story books, models, paintings, sketches, charts, graphs, clay models, cutouts, and even picturizations of rhythm and tone quality in music, were on display.

For McClusky, this kaleidoscope of child-made media, capturing a range of sensory activities and showcased in the halls of a world-class museum, offered a dazzling preview of what the media-enriched and participatory classrooms of the future could be. "The school-made visual material and the individuality shown in its use does not resemble the formal factory-made stuff that many commercial concerns are trying to foist upon the elementary school," he concluded.[32]

Over a decade later, in 1940, McClusky wrote an essay looking back on how the visual education movement had evolved in the intervening years. For him, the most significant development was a shift away from teachers and students passively consuming commercial collections of visual aids, or passing over them altogether, in favor of engaging in "active participation in the production of materials" that better suited their needs. McClusky had also shed some of his distaste for commercial technologies now that they had become affordable and accessible enough to foster more grass-roots practices of production in the schoolroom:

> While it is true that the picture and radio have become potent forces in moulding [*sic*] thought and action, it is fortunate that the ability to make fine pictures and to use the microphone is no longer limited to the chosen few. Modern inexpensive equipment makes excellent results possible to hundreds of amateurs. The tricks of the trade are no longer confined to the secret chambers of professionals. . . . Not so many years ago "educators" were clamoring for someone to make pictures for them. The old cry was, "We would use visual education, if we could buy good pictures." Now teachers and pupils are able to make their own pictures.[33]

McClusky's celebratory reports about the progress toward a more participatory, multisensory, and technology-enriched American classroom—where mediated content would be liberated from the "secret chambers of professionals," and made by the teachers and learners themselves—highlights how interwar educators simultaneously managed to push back against notions of passive media consumption while making room for more media devices in education. In this view, embracing a more grassroots, recombinant, and multisensory vision of media use in the classroom was critical not just for students' academic progress but also for their very individuality and potential to grow into democratic citizens in a mass-mediated society.

Making Homegrown Media Collections and World-Minded Citizens

As educators cautioned against overreliance on the commercial pipeline of ready-made aids, many called instead for more localized, user-driven practices of collecting, recombining, and creating heterogeneous materials for school use. In contrast to commercial advertisements for visual aids, which depicted children seated in neat rows of desks, raptly viewing screens like spectators in a theater, visual educators preferred to characterize the visual education classroom as a "laboratory" in which teachers and students actively mixed and created their own learning materials. As one director of visual education at a Chicago high school put it in 1938,

> As the cost of devices to aid the teacher becomes lower, and as more and more practical devices are being developed commercially, we find the heavy bookish atmosphere of the old-style classroom dispelled by the laboratory atmosphere, where teacher and pupil learn together by seeing, hearing, and doing things.

> Charts, pictures, slides, films, records—all these are part of the scene; and the more they are home-made by the pupil himself, the greater the interest and the value they have to the pupil.

For these educators, the richest school media collections were those compiled and controlled not by a few corporate manufacturers but rather by individual teachers and students who creatively marshaled together collections of purchased, received, gathered, homemade, and humble materials to serve their instructional needs. As a Wisconsin teacher wrote on building up a collection of visual aids in 1936, "A teacher new to the task may well ask: where does one obtain appropriate pictures? Of course, they can be purchased in assorted collections of postcards, prints, or etchings; but how much more interesting pictures are, if they have been personally collected."[34]

In the interwar years, a host of new educational media forms emerged in reflection of these ideals. School museums, realia collections, bulletin boards, scrapbooks, and opaque projection all became staples of American instruction, even as their growing presence in schools was overshadowed by popular discourses about the educational possibilities of motion pictures, radio, and other high-tech media machines. As an English teacher wrote in a 1928 essay in praise of her lowly bulletin board, "We are saying much about visual education and we are buying expensive machines for moving pictures and the like, and yet we neglect to make the most of an inexpensive but effective means which most of us have right at hand." Or as another teacher and model-making enthusiast observed in 1946, "Many of us have thought too long in terms of ready-made equipment. Now, what can we do as interested teachers and live-wire students in our own classrooms and with simple tools?"[35]

While the affordability and availability of homegrown materials was a big part of their appeal to teachers—particularly during the Great Depression when school budgets were constrained—advocates also described their homemade and recombinant attributes as more compatible with progressive ideals of active, student-centered, and world-minded learning. As Joseph Weber wrote in the *Educational Screen* in 1924,

> In the absence of motion pictures, stereographs, and lantern slides, the teacher should not become discouraged. She is really fortunate, in a sense, because then the entire class can turn sense realists and make a project of the hunt for visual materials. Success is bound to await them, for the game is plentiful in this field.

> The textbooks are filled with good illustrations, magazines like *Popular Science* or the *National Geographic* could not be richer in content, kodaks [*sic*] can be make to click unceasingly, the natural and artificial environment can be drawn on without limit, and models can be made and exhibits arranged in unending succession.[36]

Casting the quest for AV materials as a "hunt" and "game" in which students would act not as unthinking receivers but rather as active explorers and creators of their own learning devices, visual educators like Weber advanced a radically new understanding of children as mindful, discerning agents who could skillfully poach and curate educational content from a widening sea of ubiquitous, global, and ephemeral media information.[37] This approach offered an optimistic alternative to the rising moral panic—reinforced by new behavioral research such as the Payne Fund studies—that suggested children's attitudes and behaviors were easily molded by the mass media they consumed.[38] Educators' calls to foster more heterogeneous and homemade collections of media in school grew even greater in the lead-up to World War II, when concern about propaganda was rampant and standardized teaching materials quickly became obsolete amid the rapid changes in global affairs. In this climate, flexible, do-it-yourself AV collections came to be seen as beneficial not only for academic reasons but for strengthening students' information literacy and defenses against undemocratic ideology too.[39]

School Museums

These ideas began to percolate in early twentieth-century discussions about school museums—curated collections of artifacts, AV media, and texts that were arguably the first educational technologies celebrated for their multimodality. There were two types of school museums: traveling collections of objects arranged by urban museums and loaned out to surrounding schools, and in-house collections maintained and displayed by schools on their own premises. Three of the preeminent traveling school museums—the Philadelphia Commercial Museum (established in 1900), the SLEM (1904), and the Field Museum of Natural History Harris Loan Program in Chicago (1912)—consisted, at least in part, of artifacts and exhibits left over from the world's fair expositions in these cities. This fact endowed them with tremendous appeal in the eyes of educators who were interested in using technology to teach about the international world, as world's fairs were commonly understood as substitutes for international travel and eth-nological encounter.[40] The SLEM, built with materials donated to the city after the Louisiana Purchase Exposition of 1904, was the envy of many urban school systems and—along with a similar program in New York City, the School Service of the American Museum of Natural History, which launched in the same year—became a catalyst for the visual education movement nationwide.[41]

Nicknamed the "museum on wheels," the SLEM packed extensive col-lections of artifacts, natural specimens, costumes, exhibits, books, stereo-graphs, lantern slides, and motion pictures into trucks, and loaned them to Saint Louis schools on a weekly basis. Its slogan—"Bring the world to the child"—accentuated the shared mission of museums and public schools to aid in the mass enlightenment of urban, working-class, and immigrant children. A 1922 report published in *Visual Education* suggests the program was a cross between a mobile media library and cabinet of curiosities, car-rying "to the schools of St. Louis nearly 80,000 groups of museum material, ranging from a slip of race paper and a sample of linseed oil to a Navajo blanket and a stuffed eagle, from a charmingly illustrated geographical reader to a motion-picture film on the cotton industry."[42]

Proponents of school museums portrayed them as the antithesis of pas-sive education and mass entertainment. Everything in the collection was "meant to be handled" and "intended for hard use."[43] Writing in the *US Bureau of Education Bulletin* in 1924, Saint Louis assistant superintendent and SLEM director Carl Rathmann reported that like the world's fair that had given rise to it, the educational museum approximated for students the experience of visiting faraway lands. But unlike popular visual attractions, the school museum's claim to encyclopedic and experiential value lay in its diverse array of artifacts with "real" connections to people and places from around the world, each waiting to be activated by being acted *on* by inquisitive students and teachers.[44] Bringing foreign objects directly to the students' fingertips, the school museum allowed them to physically explore and virtually experience the lifeways of other peoples. Furthermore, unlike excursions to actual museums, which Rathmann warned could promote "scattered interest," and take on the feeling of a "pleasure trip" or "picnic," engaging with collections in school promoted an *educational*, not merely entertaining, sensory experience:

The illustrative material must be so used as to cultivate the child's imagination and to awaken in him a desire to learn more about the world in which he lives and to give him the power to picture to himself materials, conditions, processes, and influences which we have no means of showing him in concrete form. Merely giving him an opportunity to see and to observe the material is satisfying his curiosity, entertaining him, but not educating him.[45]

Bringing the museum to school could thus provide students with something akin to Dale's ideal of a "base" of sensory experience that could prepare them to more intelligently engage with a complex world of symbolic information beyond the schoolroom. Here students would deploy a disciplined and studious gaze, in contrast to the "excessive wonder" or "museum drunkenness" that according to some museum observers, visitors were prone to experiencing when wandering through actual museums—a distracted gaze perilously close to that of viewing a film or some other visual attraction.[46] In the eyes of advocates, school museums presented visual materials in a way that balanced wonder and delight with disciplined instruction, awakening students' interest in the subject matter while also training their sensorium.

School museum exhibits were described as interactive, encyclopedic, and utopian spaces of discovery in which students could explore and grasp relationships between a diversity of human cultures, life-forms, and geographic environments. But despite their claims to universality and realism, they were heavily constrained by the dominant representational strategies and colonial ideologies of the world's fairs and natural history museums on which they were patterned. Educational museums represented the world through familiar dichotomizations of "primitiveness" versus "progress" that were commonplace not only in museums but also in the mass cultural landscape from which schools sought to differentiate themselves. Indeed, as Rathmann put it, the goal of having students handle and compare objects from different countries was not just to become aware of their differences; it was to understand "their place among the nations" and their inhabitants' "state of civilization."[47]

Scholars have suggested that notwithstanding museums' claims to furnishing global knowledge and enlightenment, their collections are often deeply illusory, creating misrepresentations of relations between disparate objects and cultures, and arranging materials in ways that reinforce Western epistemologies of racial and cultural superiority. Traditionally, natural

history collections have elided the colonial power relations, labor, exploitation, and violence that led to their acquisition.[48] These elisions are particularly evident in the SLEM's cotton and rice exhibits, which encouraged
children to contemplate the wonders of Western industrial progress by handling raw materials extracted from distant colonies and by the cotton industry in the American south. As Rathmann described it, the cotton exhibit
gave students an imaginary tour of the international process by which this
raw resource, harvested with "primitive and crude implements used by the
inhabitants of the Philippines" as well as by "Negroes in the field" in Louisiana, could later be transformed into a valuable market commodity through
the miracle of "magnificent machinery in the large eastern factories" of
the United States.[49] This lesson, which celebrated the "successful human
genius" of capitalist technology and processes while avoiding any discussion of oppressive labor conditions or the history of slavery, likely gained
conceptual reinforcement from the assemblage of media and materials used
to teach it. Students examined samples of raw cotton and models of tools
from the Philippines with their naked eyes and bare hands. Then, they
viewed images of Black workers in the cotton fields of the American South
and cotton-processing factories in the eastern US through the mechanical
aids of the stereoscope and lantern slide. In the process, students learned
about the commodity's journey from East to West, and from crudeness to
refinement, through the recommended educational progression from "concrete" to "abstract" experience—from raw material to mechanical media.
While this lesson epitomized the era's ideals of multisensory and world-
minded learning, its ordering of objects reinforced normative and racist
ideologies that celebrated the expansion of American empire, notions of
the white, Western world as the pinnacle of civilization, and importance
of industrial development in determining the global distribution of power.

Realia

School-based collections of visual and audio media often reflected larger
tensions between emphasizing pluralism and patriotism, and nationalism
and internationalism, in the curriculum. At the same time that educators
used collections to illustrate American expansionism and progress, they
pointed to their diverse origins and content as evidence of their suitability
to teaching tolerance and the "brotherhood of man." After World War I, the
rising interest in teaching internationalism in schools led many teachers of

geography, social studies, and foreign languages to embrace the technique of teaching with realia, or things with origins in or "real" connections to a particular foreign country or language. "Realia means real things or realities," explained Amelia Milone, a teacher of Italian at Benjamin Franklin High School in New York City. "[It] covers everything that is illustrative of a nation's real life and thought—literature, history, geography, institutions, manners, and customs."[50] Collections of realia included everyday items such as postcards, posters, mounted pictures, movies, coins, restaurant menus, pamphlets, flags, ticket stubs, handicrafts, records, maps, slides, songbooks, and correspondence with students abroad. While some educators defined realia only as those artifacts with "concrete" connections to another place or culture, others used the term almost interchangeably with "visual aids" to refer to a broad range of visual, tactile, audio, and sensory materials that could illustrate daily life in distant lands. Gathered by students and teachers from immigrant relatives, international businesses, travel agencies, and pen pals, realia collections were a remarkably grassroots, global, and multimodal technology of instruction that seem to presage, in some ways, the connective, recombinant, participatory, and world-channeling features of today's internet. As one educator summarized them in 1937, "*Realia* are the classroom links between Europe and America, the stimuli to curiosity, and the sources of subtle culture."[51]

Lillian Stroebe, a professor of German at Vassar College, argued that it was only through the collection and study of these sorts of authentic cultural artifacts that students could obtain "the real knowledge of a foreign country" without traveling there on their own. She instructed foreign language teachers in training to collect realia during their university studies abroad, or gather them from books, ethnic businesses, and other outlets in their communities. But like most international materials used in American schools, realia were not neutral, and frequently served progressives' goals of teaching about the international world while promoting Americanization and patriotic nationalism. According to Milone, who incorporated Italian songs, newspapers, recordings, and motion pictures into her teaching in the era of Benito Mussolini, such activities would make "resourceful and better citizens" of her diverse student body, many of them descendants of Italian immigrants. Realia "will not only foster a spirit of sympathy, understanding, and brotherhood, but will enable our 'little citizens' to criticize more intelligently and constructively our own customs and institutions," she wrote. "Our own cherished ideals of democracy will stand forth more prominently when contrasted with the principles of other nations." Other uses of realia reinforced dominant racial and ethnic stereotypes. As Stroebe surmised, realia from Latin America were useful for raising Americans' awareness of the region as an emerging trading partner, but such materials would necessarily have less intellectual value than ones from Europe. "Some subjects . . . might be dealt with more superficially than in the case of other foreign countries, as there is little doubt that Spanish America has less to give to American students intellectually than the other foreign countries studied, however important its present or future commercial relations with the United States may be."[52]

Figure 3.2
Students learn with "dioramas, objects, models, pictures, and realia" on loan from the Department of Visual Education in Topeka, Kansas. Dorothea Pellett, "Tips from Topeka," *See and Hear* 1, no. 5 (1946): 13.

In the 1930s, homegrown AV collections became increasingly favored devices for teaching about the international world, as educators criticized textbooks and other commercially prepared aids for failing to keep abreast of the rapid changes in foreign affairs. In 1935, Marian Evans, the supervisor of visual education in the San Diego schools, wrote that teachers should begin mobilizing more flexible and ephemeral media texts, such as magazines and newspapers, to present students with not only the "panorama view of the past" but the "drama of the living present" too, including current events, scientific inventions, and social problems. She praised the efforts of the WPA's newly established Museum and Visual Aids Project, formed that year under Roosevelt, for bringing this dynamic approach to visual education to the schools. The project employed artisans and tradespeople in public schools, museums, and libraries to "create a wealth of objective and pictorial materials," including paintings, models, slides, and dioramas. This little-known effort of the WPA went by a number of names in different states, including the Museum Extension Project in Pennsylvania, the Visual Aids and Museum Assistance Program in Connecticut, and the Visual Education Project in Rhode Island.[53] For Evans, these locally sourced and federally supported collections of visual materials, made by skilled craftspeople and in accordance with local educational needs, were the epitome of a visual education program designed for informed, democratic citizenship. By enabling the curriculum "to be kept up-to-date and abreast of current issues," she wrote, students would be able to "frankly face the present controversial issues on social, economic, and political questions which they, as citizens, will be called upon to answer."[54]

Bulletin Boards, Scrapbooks, and Assemblages of Print Matter

One of the most common ways in which students encountered collections of media in school was through humble assemblages of print matter, presented via bulletin board displays, scrapbooks, and opaque projections. Compared to the prefabricated, mass-produced technologies of film, slides, radio programs, and recordings peddled in industry catalogs, these flexible, paper-based media could be easily acquired, combined, updated, and incorporated into daily instruction by teachers with little or no technical experience. Teachers prized bulletin boards and scrapbooks as flexible, collaborative, and thrifty devices that could compensate for the obsolete or otherwise-unsatisfactory content in their textbooks and other standardized teaching aids.[55] Popular uses of bulletin boards included the creation of "hot spot" maps illustrating current events and conflicts abroad, the display of world flags or scenic pictures from travel magazines, and a regularly updated visual digest of national holidays, historical figures, and local and global news.[56] A 1924 account of a bulletin board in a Rhode Island classroom highlights how educators regarded it not only as a flexible visual supplement to geography textbooks but also as a collaborative text that encouraged the development of international awareness and information literacy in children:

> In one corner of the room hangs an inexpensive bulletin board made of a bit of green or brown denim or burlap. Here are hung from day to day pictures that the pupils have brought in from the rotogravure sections of the great newspapers, clipped from magazines, or diagrams and cartoons that they have made themselves. Such scenes as the President reading his message to Congress; the trial flight of the Shenandoah; the great Japanese disaster to take at random a few current events brought in by the pupils, and edited by the teacher, to select those of real worth. Around the chalk-trays are displayed a careful selection of pictures cut from such magazines as the National Geographic or Asia, these mounted on regular sizes of cardboard. They are left, these of Egypt, the Arctic, Japan, or our own country, for a number of days, and in their leisure moments the boys and girls gather around them and absorb the colorful life of those countries which are assigned for study in their geography texts. These rooms are building up real collections of increasing value from year to year. And yet the expense of these is trifling.[57]

Teachers suggested that if managed properly, with content produced and refreshed regularly by students and teachers alike, the bulletin board could become "a live part of the classroom" and contribute to a "wide awake class" savvy in world affairs.[58] But while some educators believed that having teachers and students clip and reassemble global news and information would provide an opportunity for healthy critical thinking as well as informed citizenship, others warned against giving them too much authorial control. Bulletin boards should be presented in such a way, one Indiana educator wrote in *Social Studies* in 1938, to avoid "indoctrination of the students with the teacher's own ideas as to war and peace," and "detracting in any way from the true spirit of patriotism or national pride."[59]

One commercial device that benefited from the push for incorporating more homemade and recombinant media in the classroom was the opaque projector, which allowed teachers to instantly project any document, image, or small object onto a screen. The ability to quickly curate and project their own assemblages of content, as opposed to the more onerous process of making their own glass slides or relying on images preselected by commercial slide producers, made opaque projectors particularly appealing to teachers.[60] "One of the biggest advantages in the use of still pictures is derived from the fact that the teacher herself determines the exact order of the materials to be utilized," wrote one teacher of a class on international relations in 1945. "She can design them so as to incorporate direct teaching principles into the visual lesson. She can include the latest word, map or chart from the morning paper."[61] Producers of opaque projectors, such as the Spencer Lens Company, were keen to point out how their products kept teachers in charge of the lesson, and the screen "alive" with "up-to-the-minute" content culled from the latest newspapers and magazines.[62] During World War II, advertisements for the Spencer Delineascope opaque projector underscored its status as a cutting-edge technology of war preparedness, noting that it was used to train American troops and civilians for victory in the fast-changing global conflict. Such depictions linked the presence of user-driven, recombinant information technologies in the classroom with ideas of informed citizenship, world-minded thinking, and national security.

Clipping and creatively combining content from magazines, newspapers, books, and other printed materials was a multivalent act—one that teachers could use to both comply with dominant institutional imperatives or subtly push back against them. As Ellen Gruber Garvey has argued in her research on scrapbooking, cutting up mass-produced print matter to create something new is exemplary of the tactic of cultural resistance known as "textual poaching." Taking scissors to the "sacred" products of mass culture, textual poachers resist the recommendations of a standardized, one-size-fits-all media system by removing content from its intended context and reformulating it in a manner that better suits their needs.[63] For many educators, the practice of remixing and repurposing available media was the only way to compensate for the shortcomings of the educational media and textbook market, particularly when teaching about world affairs during wartime. As Maurice Hunt, an Ohio teacher of high school social studies, observed in 1944, by creatively drawing together a variety of films, magazines, newspapers, pamphlets, and other up-to-date media, teachers could produce rich instructional aids that overcame the "serious textbook

KEEP YOUR SCREEN _ALIVE_

Figure 3.3
Advertisements for the Spencer Delineascope opaque projector pledged to help teachers keep their screen "as alive and up-to-date as the latest newspapers and magazines." American Optical Company, *Educational Screen* 25, no. 7 (1946): 357.

weaknesses"—that is, the obsoleteness of information, superficial treatment of social and political problems, and narrow, "United States-centered" worldview—typical of many standardized instructional aids.[64] By encouraging more cosmopolitan and democratic media practices in the classroom, educators like Hunt believed they could instill in students a more cosmopolitan and democratic outlook on society.

In 1934, George Mark, a professor of education at the State Teachers College of Shippensburg, Pennsylvania, published an article in the *Junior-Senior Clearing House* to demonstrate how such a hands-on and cosmopolitan approach to new media in the classroom would bolster students' democratic sensibilities. Observing the rise of authoritarianism in Europe and noting that textbooks had failed to "foresee the social upset through which we are now passing," Mark argued that American adolescents should study— and subsequently cut up and create scrapbooks of—current newspapers to develop a deeper knowledge of global affairs. "World-minded citizenship is the American ideal," he declared, and for teaching this, standard-issue textbooks paled in comparison to students' critical engagement with up-to-date periodicals. "Probably no textbooks now in use meet present-day conditions," he wrote. "Pupils need to study what is now happening, and think more of tomorrow than of yesterday. . . . The textbook would hardly be a competitor with the newspaper in point of aliveness, directness of application, recency of events, and in rivalry of ideas." Newspapers provided raw material for students to create scrapbooks (which he called "workbooks") on major geopolitical figures and countries, such as Joseph Stalin, Mussolini, Adolf Hitler, Roosevelt, and Japan. By taking scissors to news reports of totalitarianism from abroad, students would build up their faculties of critical thinking and democratic citizenship, as each would produce something new and individualized out of the mass-produced newspaper they used. As Mark proclaimed, "There is no end to the workbook activity. It would be an enterprise of self-help, of invention, of individual selection, of personality, of eye and ear curiosity, of English composition, of research, of artistic expression." Most important, Mark noted, the process would not only teach students vital facts about contemporary geography and geopolitics but also awaken their interest in tackling world problems: "Unconsciously the pupil would be becoming world-minded; and one could believe with satisfaction that American citizens were being reared in an atmosphere of living problems, of personal zeal for the purposes of today, and it is to be hoped, with determination to be a participator in the democratic solution of our present-day difficulties."[65] Accounts like these underscored the emergent belief among American educators that mass-produced media, if collected, created, and refashioned with discernment, and while creating fascists elsewhere, could produce open-minded, democratic, and problem-solving global citizens within the classrooms of the United States.[66]

Teaching Media Criticism, Defending Democracy

The idea of using participatory, multisensory, and world-minded techniques to build healthy civic behaviors in youths was central to the early
career work of Dale in the 1930s and 1940s. Alarmed by the growing social
influence of film, radio, and newspapers in the United States, and the use
of propaganda by authoritarian regimes in Europe, Dale worried that American adolescents' uncritical consumption of media along with their schools'
traditionally passive, verbalistic approaches to instruction were leaving
them ill prepared to identify and solve the pressing social problems of their
time. He described movies, radio, and the press as the "unlicensed teachers"
and "informal school" that could weaken democracy if audiences were not
taught to question their representations and claims. In his view, schools
had an obligation to not only provide students with a broad base of participatory, "rich experiences" against which to evaluate the increasing flow
of mediated messages in modern life but also to "develop the thoughtful,
discriminating user of the mass media" through lessons in media criticism.
For Dale, learning to appraise films, radio programs, newspapers, and other
forms of mediated information according to their formal composition as
well as social messages was an exercise of good citizenship, mode of social
problem solving, and defense against undemocratic propaganda and group-
think. Dale was optimistic that teaching such skills could "raise the standard" of humankind by generating demand for films and other media that
improved understanding among nations and races.[67]

Born into a farming family in Benson, Minnesota, in 1900, Dale moved
to Rugby, North Dakota, as a child. A bright student, his interest in education stemmed from his own experiences in rural schools, where he was
often frustrated with the rote quality of instruction, and what he later called
the "significant, experiential difference between pronouncing the words
and learning their meanings." After graduating from high school at age fifteen, he completed his bachelor's degree in education from the University
of North Dakota via correspondence—an experience that was formative to
his later interests in instruction and new modalities of communication.
He worked as a teacher and superintendent in schools in rural Minnesota
and North Dakota, where in 1919, he encountered his first modern visual
aid in the classroom: a thirty-five-millimeter industrial film that had been
donated to the school. In a testament to the ad hoc nature of technology in

schools at that time, Dale later recalled that he simply "didn't know what to do with it."[68]

In 1924, Dale took a job as a junior high school teacher in the progressive school system of Winnetka, Illinois, a growing suburb of Chicago. His arrival coincided with the tenure of Superintendent Carleton Washburne, an influential school reformer and Dewey disciple. Washburne was in the midst of implementing his "Winnetka Plan," a curricular initiative aimed at educating the "whole child" through a combination of creative, social, and individualized activities. The schools had also just adopted the controversial social studies pamphlets of Harold Rugg, a former professor at the University of Chicago who had recently moved to Columbia University's Teachers College. Rugg, whom Washburne brought to Winnetka as a guest speaker, envisioned social studies as a unified course of academic study that would integrate multiple subjects, including geography, history, and civics, to help junior and high school students understand the complexities of "present living." Like many white progressives, Rugg was an internationalist who held a sympathetic view of European immigration, but largely ignored matters of segregation and structural racism against African Americans, Asians, and other nonwhite groups. His vision for social studies included an emphasis on teaching new information literacies. He urged teachers to have their students make scrapbooks of "newspaper-clippings, articles, cartoons, [and] statistics" on world affairs, and curriculum designers to consider "what historical allusions are needed by all people who read newspapers and magazines, and what were the great contemporary problems and issues which all children should have an appreciation of."[69] This progressive and "free-thinking climate" at Winnetka inspired Dale to enroll at the University of Chicago to pursue a doctorate in education in 1925.[70]

At Chicago, Dale was surrounded by a number of emerging leaders in the visual education movement. Frank Freeman, a professor of educational psychology, had just published a landmark study showing the limited effectiveness in instruction of visual aids, including motion pictures, lantern slides, and illustrated texts. In it, he cautioned against privileging the film above other aids, the dangers of "passive receptivity," and overlooking the importance of the teacher in determining the efficacy of visual materials.[71] Dale studied under Charles Judd, the head of the NEA's newly formed "Judd Committee," which was in the midst of forging a partnership between the educational community and film industry to study the instructional potentials of film. He also studied under Werrett Wallace (W. W.) Charters, a former student of Dewey's.[72] Charters, who became one of Dale's closest mentors, moved to Columbus in 1928 to become director of the Bureau of Educational Research at Ohio State University.

That year, after finishing his dissertation on curriculum design, Dale tried his hand at working in the educational film industry. After a brief stint of writing teaching guides for Eastman Teaching Films, a subsidiary of the Eastman Kodak Company, he accepted an invitation from Charters to join him at Ohio State, where he remained as a professor for the rest of his career. There they embarked on several collaborative initiatives in visual and audio education research supported by the Payne Fund, a philanthropic foundation established in 1925 by Frances Payne Bolton, a member of a prominent Ohio political family and advocate for child welfare. Their first project was an evaluation of the Ohio School of the Air, the educational radio program established at Ohio State and broadcast over WLW Cincinnati in early 1929.[73] The second project was a large-scale study of the effects of motion pictures on youths—the famous Payne Fund studies. Dale wrote two of the eleven studies in the series, titled *The Content of Motion Pictures* and *Children's Attendance at Motion Pictures*. His third contribution to the series, a textbook for high school students called *How to Appreciate Motion Pictures*, became the only volume of the series to exceed sales expectations and propelled him to the forefront of an emerging movement to teach film appreciation in schools.[74]

Dale's emergence on the national stage was preceded by years of mounting concern among educators and social reformers about young people and media reception. In the 1910s and 1920s, a range of new and popular media—including movies, dime novels, and comics—drew criticism from educational, religious, and civic groups for absorbing youths' leisure time, and negatively affecting their morality, attitudes, and behaviors.[75] The prevailing view within the visual education movement was that the presence and persuasive influence of mass media—particularly movies and radio—were only going to increase with time, making it necessary to either reform the way in which the media industries produced their products or how children consumed them. The 1920s and early 1930s saw more of the former approach. Reformers called for censorship of the film industry, establishing local and regional censorship boards to edit as well as ban certain films from theaters. They also tried, unsuccessfully, to ensure an educational role for the radio broadcasting system—an effort that was largely defeated with the passage of the Communications Act of 1934.[76] Commercial radio quickly crowded out independent broadcasters as a result of the law, leading a formerly vibrant landscape of educational radio stations, numbering over 120 in 1925, to shrink to only 36 stations by 1935.[77]

By the mid-1930s, educators began to shift their energies from reforming the site of media production to that of reception, focusing less on regulating the media industries, and more on transforming the public into more critical and discerning consumers of the existing media culture.[78] A pivotal moment, according to John Nichols, was the release of the Payne Fund studies. The first eleven studies in the series, along with a popular summary, *Our Movie Made Children* by Henry James Foreman, provided social scientific reinforcement for prevailing assumptions that motion pictures had a powerful, potentially harmful effect over children's attitudes and behaviors, and therefore needed to be more tightly regulated or reformed. But the final volume in the series—Dale's *How to Appreciate Motion Pictures*—presented educators with an alternative way of thinking about the instructional potentials of motion pictures, and critical media reception and citizenship more broadly.[79]

Believing that teaching the public to intelligently view movies was more important than reforming the industry, Dale wrote *How to Appreciate Motion Pictures* as a textbook for high school students. Its goal was to teach teenagers to "judge" popular films on the basis of both their formal composition and social messages, with the ultimate aim of improving the motion picture landscape by elevating audiences' tastes and generating demand for higher-quality films. Whereas visual educators had previously tended to categorize films as either appropriate or inappropriate for young people, and discouraged them from viewing the latter, Dale instead *encouraged* students to view controversial films—particularly ones dealing with issues of prejudice, war, and crime—to unpack their moral messages and discuss their implications for society. His textbook combined activities in artistic and technical analysis, including suggestions for evaluating dialogue, acting, plot, and cinematography, with discussion questions designed to prompt students to reflect on the film's treatment of social problems. On the issue of racial and ethnic stereotypes, for example, Dale asked students to identify instances in which films tried to get laughs from the audience through exaggerated depictions of foreigners and minorities. "One must inquire whether the humor thus gained was in good taste," he wrote. "Was it vulgar? Was it secured at the expense of some particular race or group of people?" The implication was that if young people could be taught to detect the presence of stereotypes, prejudiced viewpoints, or other antidemocratic representations in film, they would then soundly reject those kinds of pictures at the box office in favor of more prosocial films. As Dale later proposed in 1938,

> What practical steps can we take to control, in some measure, the depictions of other races and nationalities on screen so as to further, rather than retard, harmonious relations with other peoples? First, we must make motion pictures unprofitable which show these stereotypes, and make those movies which are honest and accurate in their portrayals of race and nationality relationships successful at the box office. This means, frankly, the organization of the movie consumer.[80]

Dale's interest in eliminating negative stereotypes through the "organization of the movie consumer" emerged from one of his other, less acknowledged contributions to the Payne Fund series: a study of fifteen hundred feature films called *The Content of Motion Pictures*. Attempting to identify prevailing themes and social values in popular movies through a study of their plot summaries, the research was rudimentary but nevertheless revealing.[81] Looking at theatrical movies produced in 1920, 1925, and 1930, Dale noticed patterns that troubled him, particularly in the "depiction of foreigners and less-favored races." He found that a third of all films were set outside the United States, which suggested to him that movies were, for the average citizen, an important source of information about the international world. Studying a subset of forty films in detail, he discovered that foreigners and minorities were often portrayed in an exaggerated or "unattractive" manner to "furnish comedy relief," and that the "Negro in the movies is, almost without exception, either an entertainer or a servant of white folks." Citing the newly published research of social worker Bruno Lasker on the formation of racial attitudes in children (discussed in greater detail in the following chapter), Dale worried that this tendency of the movie industry to represent nonwhite racial groups "by their worst representatives" posed a direct threat to the progressive educational goal of "breaking down of racial prejudice and setting up of intelligent attitudes towards persons of other races." It was here that Dale first developed his approach to reforming society by educating audiences to reform the movie market, suggesting that informed consumers would want to withhold their support for films that contained negative stereotypes. The discerning parent, he wrote, would "avoid sending his children to [films] which are patently prejudicial," and "make use of those motion pictures which give an honest understanding of foreign persons and underprivileged groups."[82]

Dale's vision of combating the social inadequacies of motion pictures through more mindful consumption was groundbreaking in that it placed the power for meaning making and representational change in the hands of citizens rather than the movie censor. But it was also problematic in several ways. First, it required the public to see and patronize films in order to judge them, thus limiting the financial impact of their subsequent decisions to accept or reject certain films. Also, it centered on the idealistic assumption that viewers would *want* to reject films that contained stereotypes once they were taught to detect them. Dale's ideal of promoting interracial and international harmony through media criticism did not engage with the underlying histories, ideologies, or active policies of legalized racism, segregation, immigrant exclusion, and discrimination that made such characterizations ubiquitous in American popular culture as well as culturally resonant with white audiences in the first place. Finally, as Dale himself admitted, the box office was a flawed instrument for registering citizens' preferences, as African Americans were barred from attendance at many movie theaters.[83]

Dale's concern for teaching students to identify media stereotypes only intensified in the latter half of the 1930s, as he witnessed Hitler's rise through the elaborate use of movies and other forms of propaganda. In 1936, on a trip to Europe as a motion picture specialist for the Child Welfare Committee of the League of Nations, the first of many international advisory roles in AV education that he would hold in his career, Dale visited what he referred to as the "German Film Institute" in Germany. He left in awe of the Germans' far-reaching use of film and other media for instructional purposes, conceding that their main purpose was not really instruction but instead indoctrination. As he would later rue during the war, "The Nazis had already learned how to use the motion picture as an agency of propaganda and information, something we still are learning today." Dale wrote this as he worked for the US Office of the Coordinator of Government Films, an agency created during wartime to help orchestrate the production of American propaganda films for the war effort. Because of that visit to Germany, "I realized how slow the democracies have been to put to work the instruments of mass communication. I say put them to work. We have already used them for play. . . . We are realizing today as never before, that the motion picture and visual materials are weapons of war as well as weapons of peace."[84] Frustrated by the lag of the American film industry and government in producing films that aided the cause of education and democracy, Dale doubled down on his efforts to reform the "uncritical attitude" of the American consumer. "This uncritical attitude has become the working ground of the propagandist in many countries today," he warned in a 1937 speech before the Williamstown Institute of Human Relations at Williams College. "With the motion picture and the radio at his command, the propagandist can, through the use of selected narrative and drama, convert the mist of popular prejudice constantly hovering about into a storm of blind antagonism." Dale reiterated his belief that the surest way to fortify democracy, teach tolerance, and elevate the American media landscape was to "inoculate the American public with the vaccine of critical judgment."[85]

By this time, Dale was beginning to speak of teaching media criticism in reference to a multitude of media and information technologies, no longer just motion pictures. As Acland has noted, Dale's "multimedia ambition was amply evident" in his work as early as 1935, when his colleague Charters secured money from the Payne Fund to launch a Radio-Press-Cinema Project at the Bureau of Educational Research, building on the success of the motion picture studies. At this time, a "radio appreciation" movement was gaining momentum across the country, mirroring both the film appreciation movement and what David Goodman has called the distinctively "civic ambition" of the American broadcasting field in the 1930s.[86] In junior high and high schools, students formed radio listening clubs, took field trips to local radio stations, wrote critical reviews of popular programs, staged their own "mock broadcasts" of radio dramas and town halls of the air, and created their own live "broadcasts" over their schools' public address systems.[87] For radio appreciation advocates, students' active and critical participation in radio culture was a kind of training for future citizenship and informed opinion making. Dale's colleague and collaborator at the Bureau of Educational Research, I. Keith Tyler, was a leading figure in this movement, authoring *How to Judge a Radio Program* (1936) and directing *High-School Students Talk It Over* (1937), an Ohio School of the Air program that featured teenagers' live discussions about war, motion pictures, and social issues.[88]

By the late 1930s and early 1940s, Dale's multimedia vision of media literacy was evident in his descriptions of the ideal student as a "critical

Figure 3.4
"A broadcast originating in and heard by the entire school at Whitefish Bay, Wis., by means of a central sound system installation." Dale, *Audio-Visual Methods in Teaching*, 264.

viewer, listener, and reader." He likewise referred to the movies, radio, and press together as the "unlicensed teachers" that needed to be kept in check by a rising generation of discerning, media-savvy citizen-consumers.[89] With the film and radio appreciation movements underway, Dale turned his attention to print journalism, publishing *How to Read a Newspaper* (1941), a textbook similar in approach to *How to Appreciate Motion Pictures*. Addressing high school students as "news-readers—architects of an American future," the book aimed to train teenagers to evaluate both the technical and social aspects of journalism, with the ultimate goal of improving the news industry and strengthening democracy by generating greater demand for less biased, higher-quality periodicals.[90] The book urged students to read multiple newspapers regularly and appraise their components, including headlines, sources, statistics, word choices, photographs, and political cartoons, with an eye toward how such devices shaped public opinion. In a voice that reflected the urgency of wartime, Dale described the newspaper as "democracy's textbook," and a testament to the opposing value systems of democratic and totalitarian societies. "In a democracy the people read the newspapers' viewpoint and then decide for themselves," he wrote. "In a dictatorship there is room for only one voice—one will—that of the dictator."[91] Illustrated with reproductions of major newspapers' front-page coverage of the bombing of Pearl Harbor, *How to Read a Newspaper* treated critical news reading and consumption practices as a matter of not only improving the marketplace for news in the United States but also exercising and protecting Americans' democratic freedoms against the threat of totalitarianism.

Dale's view of media criticism as a democratic defense gained greater definition—and a larger platform—through his involvement with the propaganda analysis movement. In 1937, he joined the advisory board of the Institute for Propaganda Analysis (IPA), a nonprofit based in New York City and established that year by Clyde Miller, a muckraking journalist and faculty member at Teachers College, along with Kirtley Mather, a Harvard geologist and adult education advocate, and Edward Filene, a liberal philanthropist, department store owner, and son of a Prussian Jewish émigré. Other advisers included William Heard Kilpatrick, a progressive pedagogue and Dewey student, and prominent social psychologists and sociologists working in the emerging fields of communication, propaganda, and public opinion research, including Hadley Cantril, Robert Lynd, and Leonard Doob. With a mission of helping "the intelligent citizen detect and analyze propaganda," the IPA distributed study guides and a monthly bulletin, *Propaganda Analysis*, to some 3,000 high schools and 2,000 adult education groups between 1937 and 1941.[92] By 1939, the materials were being actively used in 550 high schools and colleges. The materials featured activities and discussion questions to help students analyze the persuasive appeals embedded in a wide range of communications, including movies, radio, billboard and print advertisements, newspapers, newsreels, speeches, pamphlets, and textbooks.[93] The IPA encouraged students to identify what it called the "seven devices of propaganda"—name calling, glittering generalities, transfer, testimonial, plain folks, card stacking, and bandwagon—a list that Dale applauded as particularly accessible and comprehensible to adolescents.[94] In *Propaganda: How to Recognize it and Deal with It* (1938), a study unit for junior high and high schoolers, the IPA encouraged students to become voracious and critical consumers of mediated information—watching movies and newsreels, listening to radio programs, and making scrapbooks of the print advertisements and news articles that they encountered in their daily lives. The group urged students to systematically note the appeals such communications made (for example, to people's fears or prejudices, or their desires for beauty, popularity, health, and economy), the evidence and methods they used, the audiences they targeted, and the interest groups they served. Due to this cross-media and critical approach, the IPA has been described as one of the first organizations to promote critical media literacy education in the United States.[95]

The IPA defined propaganda in a variety of ways, from any communication that "attempts to persuade [citizens] to do something they might not do if they were given all the facts" to a "means of social control or a method by which an individual or group works toward his or their own interest."[96] In its school materials, the IPA warned educators that propaganda was "assailing Americans today" from all directions, and was used not only by foreign despots like Hitler but also by the American government, political candidates, advertisers, and the burgeoning field of public relations. Despite this alarming rhetoric, the institute maintained that propaganda was not inherently dangerous but instead "may be good or bad, depending on the social point of view of the person judging it." The primary danger, the IPA wrote in its materials for high schoolers, was when propaganda reached an "uncritical audience," which "opens the way to a fascist demagogue."[97] Thus, similar to Dale's work on film and newspaper appreciation, the IPA maintained that teaching Americans to understand and actively critique their media, rather than attempting to censor or regulate it at its source, was the key to ensuring that it would not undermine democracy. "In a democracy, freedom of speech necessarily means freedom to propagandize," wrote Miller in 1938. "And this freedom implies an obligation resting upon citizens to analyze propaganda affecting their interests, and the interests of the community."[98]

Dale and the IPA's writings reveal how the budding educational ideal of a multimedia-literate, empowered public was in problematic alignment with assumptions about the supremacy of an unregulated, market-based media landscape driven by consumer interest and demand. The IPA presented its techniques as an objective "scientific method," maintaining that citizens' rational analysis and open discourse were the best instruments for rooting out falsehoods from the information landscape. While a free society supported all manner of media and propaganda, it was up to the schools to create the "intelligent, informed citizenry" that would keep such a system in check. Yet for all its emphasis on investing ordinary citizens with the power to determine the ultimate social influence of mass communication, the IPA assumed that informed responses to propaganda would be fairly homogeneous. Comprised entirely of white men, the organization did not acknowledge the diversity of viewpoints, experiences, and social positions that citizens would bring to bear on their interpretations of messages. While urging high school students to be particularly on guard against messages that fomented intolerance of particular racial or ethnic groups, the IPA did not go so far as to address the structural barriers and policies that actively prevented minorities, women, immigrants, and other marginalized people from fully participating in the information consumption practices and deliberative debates that the IPA saw as so pivotal to ferreting out the "good" propaganda from the "bad." Quite the opposite, the IPA maintained a rosy view of the American public sphere that assumed it was already fully equipped for the participation of all groups. As Miller remarked in a speech before the National Council of the Teachers of English in 1938,

> The Institute for Propaganda Analysis has a major assumption—namely, that the democratic way of life, whose criteria are set forth in the Bill of Rights, is the most desirable way of life. Here opinions are respected no matter whence they come— whether from Jews or Catholics or Negroes or foreign born. These opinions, by this assumption, must run the gauntlet of criticism just as a motor car must go through all sorts of trying tests.[99]

The United States' entry into the war in 1941 dealt a blow to the propaganda analysis movement, as patriotic viewing and unquestioned loyalty to the nation superseded interests in critical information literacy. The IPA suspended its work that year amid charges that teaching the public to question information could "hurt defense" and be seen as "unpatriotic."[100] Some observers voiced additional concerns that teaching propaganda analysis in schools made young people so skeptical of new information, they were prone to "disillusionment" along with "easy cynicism and smartness."[101] Though it resumed operations in 1946, the organization was subsequently rocked by allegations from the House Committee on Un-American Activities that Miller was a Communist and formally disbanded in 1950.[102]

Despite its brief and controversial existence, the IPA left a mark on fledgling efforts to teach critical media literacy in schools. During and after the war, a number of high school and university educators turned to IPA materials to engage with foreign as well as domestic propaganda in their lessons. Even as Dale and other AV experts joined the war effort and helped to produce propaganda films for the American government, propaganda analysis continued to be taught in several schools as one of many wartime curricular activities intended to promote critical thinking, antiprejudice, democratic attitudes, and the peaceful resolution of social problems.[103] Teachers, university professors, and leaders of student organizations published essays in teaching journals about how they deployed propaganda analysis to foster discussions on intergroup tolerance, gender, class, consumerism, and war. Amid growing interest in intercultural education, some teachers cited propaganda analysis as a technique that could help students identify where and how they acquired stereotypes about ethnic and racial others.[104] Overall, the propaganda analysis movement helped advance a more systematic and cross-media approach to teaching media criticism than had been forwarded by the earlier efforts in film, radio, and newspaper appreciation. It also strengthened the association between the critical analysis of media and liberal values of civic participation, individualism, and the defense of democratic freedoms from fascism.

By participating in the somewhat-subversive propaganda analysis movement while developing his own mainstream materials in teacher training and AV instruction, Dale helped bring into being several disparate but interrelated strands of emergent media literacy and multimediated pedagogy.[105] From his early interest in educational film in the late 1920s, he developed an increasingly diversified and participatory view of learning through media in the 1930s and early 1940s—a sensibility that culminated in his influential cone of experience in 1946. Witnessing firsthand the persuasive power of fascist propaganda in Germany, Dale came to associate the intentional and critical usage of multiple media devices in school with responsible citizenship in an open democracy. Promoting the cosmopolitan and liberal use of media in school as well as daily life as a basis for cosmopolitan and liberal citizenship, he was optimistic that more "integrated" and harmonious uses of mediated information in the classroom would lead to a more integrated and harmonious society.

For Dale, the American school's obligation in the age of fascism, mass amusements, and propaganda was to prepare the young person for informed citizenship by providing them with rich, multisensory experiences and analytic skills for a lifetime of discerning media consumption. Open-minded, critical, rational, and free to express their values, whether by voicing their opinion or "shopping" for their sources of information and entertainment, the discriminating American media user was the opposite of the mechanistic, intolerant member of a fascist or closed society. Dale laid out this contrast in a 1939 essay titled "Sponge-Minded or Critical?" in the journal *Social Progress*. Unlike those unreflexive consumers of media, whose minds, "porous as a sponge . . . absorb but do not assimilate" information,

> the critically minded are active, not passive, in their reception of the printed and spoken word or the motion picture. They constantly ask: "Is it true? Where's your evidence?" and "What do you mean by 'true'"? . . . In the school, they give the ill-informed and inaccurate teacher many an unhappy moment, but they are our only hope for progress. . . . If we can develop a generation of young people who will think clearly and feel deeply about the present-day problems, we need have no fear for the future of America.[106]

In all, Dale's interwar vision for teaching media literacy was forward thinking in its insistence that media were multiple and interlinked, fraught with bias, and requiring of reform through citizen critique. But it was also shaped by a number of problematic assumptions: that media representations should be valued or spurned according to their level of accuracy, that racial and ethnic stereotypes in media should be identified and plainly rejected, rather than interrogated or traced to deeper systemic inequities, and that liberal, civic-minded, and conscientious media consumption, as opposed to regulation of the media industries, was the way to ensure that democracy and an expanding mass media culture could coexist in harmony. As a leader in the twentieth-century movement to anchor mediated instruction in a progressive and student-centered curriculum, Dale's approach to media literacy encouraged young people to identify and reject antidemocratic appeals and falsehoods, but it steered clear of urging them to explore or reform the underlying ideologies or social structures that produced such representations in the first place.

4 Pageants of Peace: Youth Performances of Patriotism and Pluralism

World brotherhood is but the expansion of American faith.
—Ella Lyman Cabot et al., *A Course in Citizenship*, 1914

In early 1939, the educational journal the *Clearing House* reported that a "miniature world war" had erupted in the ninth-grade classroom of a New Jersey teacher named Adeline Mair. As the specter of real war loomed in Europe, Mair was one of many American educators to report that ethnic tensions were simmering in her own school, in which "thirteen different nationalities and as many heterogeneous backgrounds were rubbing elbows with one another daily." Using stereotypical language to describe the discord, she cited feuds between a "particularly combustible little Cuban girl" and a Russian Jew, a "pugnacious Irish lad" and a "would-be orator from the land of Demosthenes," and Italian and German boys who argued "bitterly over the relative merits of Hitler and Mussolini." To quell this "class of Babel," Mair turned to a trusted technique of progressive education—a pageant—to foster unity among the group.

Guided by an emerging approach to intercultural education that called for celebrating the "cultural gifts" of immigrants, she instructed students to research and display the "contribution each nationality represented in the class had made to modern civilization, and the vast importance of each one in the amalgamated scheme of American life." In a pageant before their peers and parents, students acted out the cultural achievements of their ancestors, including Russian dances, Italian painting, and German classical music, all under the orchestration of a "little fair-haired girl [who] represented America." To underscore how these different cultures were embraced by the United States, the pageant was performed underneath a large, student-made "frieze showing America standing in the middle with outstretched arms, and on either side figures representing the different nationalities and the contribution that each one has made to our country." For Mair, this "unit on foreign cultures" was a rousing success in conveying not only the value of cultural differences but also the uniquely democratic spirit of America—a lesson sorely needed as the nation faced economic depression and the likelihood of intervening in another world war. Through their cooperative effort to put on the pageant, students soon "forgot their petty differences," and "discovered the necessity of working together and of helping one another if we were to succeed in accomplishing one common objective."[1]

In the first decades of the twentieth century, visual aids were just one part of a broader assemblage of new communication tools, techniques, and practices that teachers experimented with in an effort to make students at once more world-minded and more American. As educators wrestled with the challenges of preparing a diverse population of young people for citizenship in a society transformed by immigration, industrialization, and global war, many embraced pageants, skits, and school assemblies as uniquely accessible, malleable, and participatory devices of instruction for conveying civic as well as social values. Though pageants initially gained popularity as a medium for inculcating immigrants with mainstream American values, celebrating Anglo-American history, and promoting patriotism before and during the First World War, educators later adapted them to the more complex interwar projects of teaching international understanding as a means of war prevention, celebrating the United States' immigrant heritage, and addressing race relations.[2] Armed with new social scientific research that suggested that teaching world-mindedness depended on rooting out local prejudices, educators embraced pageants and other school-based cultural performances as a way to promote common experiences and a "sympathetic understanding" among diverse communities, and "counteract" the racial and ethnic stereotypes that circulated in mass culture and society. As activities that hinged on students' participation and performance before publics of peers and parents, they figured children and youths as both the primary target and instrument of a wider social reform. In contrast to the vast, impersonal industrial and political forces that seemed to exacerbate prejudices and threaten democracy, school performances were local, cooperative, and emotional exercises that could instill in children as well as their parents and wider communities the importance of "getting together."

Pageants are overlooked devices in the history of American educational technology, but they are critical to understanding how schools and community institutions have long turned to local people and resources to transmit, through everyday acts of performance and cultural production, imagery and ideology about the nation and world. This chapter examines three practices of youth performance that emerged in public schools before World War II, considering each as a potent type of school media. It considers pre–World War I pageants as a form of mass civic instruction and Americanization, post–World War I school skits and activities as demonstrations of everyday internationalism, and the interwar high school assemblies and radio series of Rachel Davis DuBois as dynamic cultural productions that paved the way for intercultural lessons to enter into mainstream education during World War II. These varied cultural productions in schools constituted significant and fraught moments of participatory global and intercultural communication in American education and everyday life. By encouraging young people to perform and view a mix of racial and ethnic identities and signifiers in school as well as in their communities, they forged new, grassroots practices of representing difference while simultaneously centering whiteness as the dominant ideal of citizenship and positioning the United States—the "melting pot of nations"—as the international model of pluralistic democracy. Additionally, the social impact of such performances was limited by their focus on reforming social attitudes, not discriminatory laws or policies. As one interwar educator complained of schools' reliance on peace pageants for "reducing racial and religious friction," such productions gave "liberals a pleasant thrill" while ignoring the glaring realities of institutional racism and segregation within American schools and society.[3] Built on the belief that entrenched interracial and interethnic tensions could be overcome by creating positive images, feelings, and associations, school pageants taught a generalized "sympathetic appreciation" for differences while obscuring the specificities of structural inequality.

The "Hundred-Headed Teacher": Pageants as Prewar Devices of Mass Civic Instruction

The ideal of teaching youths to practice and perform tolerant attitudes in their communities took root in early twentieth-century efforts to broaden the social function of the public school. With the surge of immigration and urban populations at the turn of the twentieth century, educational reformers began to envision the neighborhood school as not only an institution of instruction but also democratic communication and social reform. Concerned that the forces of immigration, mass communication, and industrial modernity were disrupting social cohesion, reformers called for transforming schools into "social centers" to revitalize communities, serve as "wellsprings of democratic living and behavior," and promote a "more powerful form of social life."[4] The social center movement, as it became known, gained momentum in the 1910s with the passage of "wider use" laws in several states, which allowed public schools to be used for a range of community functions and services, including neighborhood meetings, playground recreation programs, motion picture screenings, and adult education in civics and English.[5] In an address titled "The School as Social Center," delivered in 1902 to the National Council of Education, John Dewey argued that more than merely educating children, the twentieth-century public school needed to be brought "completely into the current social life," and "operate as a center of life for all ages and classes." For Dewey, the decline of traditional institutions of socialization, particularly the church and family, coupled with rapid advancements in communication and transport, was leading to an increasing "commingling of classes and races" that produced "instability" as well as an opportunity to build a more vibrant democracy. In this age of waning tradition and increasing intercultural contact, the school needed to move beyond teaching traditional notions of citizenship, defined as voting or political action, in favor of a "broadened" definition that emphasized "relationships of all sorts that are involved in membership in a community." As Dewey put it, "We find that most of our pressing political problems cannot be solved by special measures of legislation or executive activity; but only by the promotion of common sympathies and a common understanding."[6]

The idea that schools could engineer cooperation among Americans of diverse cultural backgrounds became the cornerstone of the "community civics" education movement, which the NEA formally endorsed in 1915. In the late nineteenth century, civics education was limited to secondary schools, and focused on matters of political participation, the workings of government, and individual rights. The new civics, in contrast, extended citizenship education to young children and the domain of everyday life, redefining good citizenship not in terms of political action but rather cooperative membership in a community. As Julie Reuben writes, community civics education promoted a "broadened" and more inclusive definition of citizenship while separating it from the right to vote. Emphasizing community and cooperation over individual rights, progressive reformers could build support for their visions of a strong central government and social welfare programs in a way that worked around the disenfranchisement of minorities, women, and children. In this view, notes Reuben, "All could be citizens; only some could fulfill their citizenship responsibilities in the public sphere."[7]

The new civics was initially local in orientation, focusing on the "communities" of the home, city, state, and nation.[8] But after the First World War, as progressives rallied around the idea of a more active role for the United States in world affairs, civic education programs expanded to include an emphasis on young people's responsibilities to the "world community." Arthur William Dunn, an Indiana educator, served as a specialist in civic education at the US Bureau of Education during the war and has been called the father of community civics. After the war, he wrote that teaching children to adopt cooperative behaviors in their local environment could be a stepping-stone toward developing a "sympathetic understanding" of people of other nations. Furthermore, he believed the war had demonstrated the duty of the American public school system to promote democracy not only at home but also around the globe: "To educate citizens who will make democracy safe for the world is the *one essential function* of the American public school." Dunn carefully situated his call for an internationalist education in the framework of patriotic nationalism, citing that the United States, with its democratic government and "melting pot" makeup, was uniquely equipped to show the rest of the world how people of many nations could live together in peace. "The first and best service that a nation can perform for the world community is to be loyal to *American ideals*, which are becoming the ideals of an ever-increasing part of the world's population," he wrote in 1920. "The new type of patriot no longer cries, 'My country against the world,' but 'My country for the world.'"[9]

A popular medium through which schools transmitted lessons in the new civics to children and their communities was the pageant. Pageants were amateur plays, parades, and historical reenactments performed for communities by their own members. At the turn of the century, they were organized by a variety of civic and educational institutions, including churches, schools, and women's clubs, and were a regular feature on the chautauqua circuit.[10] Inspired by the patriotic festivals held across the country in celebration of the 1876 centennial, a "new pageantry" movement emerged in the first decades of the twentieth century as cities, towns, and school systems transformed by immigration and industrial growth looked for ways to solidify civic unity through community displays of local history and national progress. As David Glassberg observes, pageants represented a "powerful medium of mass persuasion capable of molding and remolding the collective identity and personality," and "could revitalize civic life by gathering citizens together in joyous play."[11] Wearing home-made costumes, and assembling in public parks, school auditoriums, and even football stadiums for elaborate productions that often lasted one to two hours, townspeople dramatized the dominant Anglo-Protestant narratives of American history as well as, in some cases, the "recent contributions" of immigrants to the local community.[12] Stock characters included European explorers and settlers, pilgrims, Indians, the founding fathers, Uncle Sam, and Columbia—also known as Lady Liberty—the white, female personification of America. Women and girls, dressed in classical white robes, performed "spirit dances" to represent progressivist virtues, such as peace, education, and charity. Described by one Indiana high school history teacher as an "idealized community epic," and the "cleanest and most wholesome form of drama," the pageant countered cultural anxieties about ethnic heterogeneity, industrial alienation, and the "mob mind" with grassroots communal rituals that glorified civic pride, social progress, and a shared understanding of history.[13]

But despite their local flavor, pageants were also ambitiously global texts that asserted the universalism of American democratic principles in the neighborhood, nation, and world. As Ralph Davol, an experienced pageant planner, wrote in *A Handbook of American Pageantry* in 1914, "No pageant can be purely parochial—always there is the universal in the local—each event is charged with meaning that touches world forces."[14]

Educators in particular embraced pageants as accessible, flexible, and participatory devices of visual instruction that could efficiently convey social messages to large groups of learners. Proponents described them as a thoroughly modern instrument of mass communication and social reform—a "living picture," "living museum," and "hundred-headed teacher" that provided uplifting educational experiences that could counter the cheap, corrupting moral influence of cinema. "When we come to compare the motion picture with the pageant picture the difference is in favor of the latter, as much as substance is superior to shadow," wrote Davol. "The human touch can never be supplanted by machine-made films." For him, pageants were a "nursery of patriotism" and "municipal invigorant" that could "impart, especially to the citizens of newer nations with which America has blood connections, the ideals of democracy and the stepping stones of history, better than the combined efforts of school teachers can do in years by customary text-book methods. The whole municipality goes to school together."[15] Indeed, advocates often claimed that the uniquely communitarian nature of the pageant made it a form of expression ideally suited to America's role as "teacher of self-government" to new immigrants and foreign nations alike. According to the educational reformer John Collier, pageantry was "the form of art which comes nearest to expressing the new social idea . . . which is destined to make the world over during the next century or so": "freedom through cooperation."[16]

For school administrators and teachers, assigning students to put on a pageant fulfilled multiple objectives of progressive education. Researching and dramatizing lessons from history, the new civics, and social studies, children engaged in "learning by doing," connected their schoolwork to the world outside the classroom, and modeled citizenship for the wider community—all while advertising the worth of the public school. As educational researchers at Indiana University observed in a 1929 study of efforts to teach international understanding in schools, "One of the effective features of pageants, bazaars, and plays, presented by school children, is that the adults will attend these activities and they receive from them broader conceptions of their duties and obligations to mankind."[17] Visual educators, who advocated for incorporating motion pictures, slides, and other sensory media into instruction, frequently mentioned pageants alongside these technological aids, describing them all as "devices" that "vitalized" learning in a way that passive, rote instructional methods failed to do.[18] Pageants could produce "stirring" experiences and lasting "impressions" in the minds and hearts of youths, while educating their unschooled parents with simple narratives and symbolism.

School pageants were deemed especially useful devices for the challenging task of acknowledging America's cultural diversity while presenting a program of assimilation. In *Schools of To-Morrow* (1915), Dewey and his daughter Evelyn surveyed progressive schools across the country. They wrote that a public school pageant about Christopher Columbus, performed in a diverse immigrant enclave of Chicago, provided a "unifying experience for a foreign community," and "very good picture of the outline of our history and the spirit of the country."[19] But as idealized representations of history and collective identity, pageants were also fraught documents of the tensions between teaching patriotism and pluralism, and nationalism and internationalism, in American schools. Amid the heightened anti-immigrant backlash of the 1910s and 1920s, hereditary and nativist groups called on schools to teach "one hundred per cent Americanism" through a curricular focus on Anglo-Saxon US history. Progressive schoolteachers and urban settlement house workers, on the other hand, urged a "gentler form of Americanization" that recognized some foreign traditions and folkways as part of the American story.[20] While the aftermath of the First World War prompted many school systems across the United States to introduce units on international understanding, provincial school boards and patriotic societies, such as the Daughters of the American Revolution and American Legion, exercised heightened vigilance over teachers and textbooks that promulgated ideas that they deemed "un-American."[21] Navigating these competing political pressures, school pageants often featured carefully selected displays of internationalism that culminated in unifying messages about American patriotism, loyalty, and universalism. By producing and participating in these productions, white teachers and students developed modalities of playfully exploring cultural differences, and expressing a generalized appreciation for other nations and races, all while performing their allegiance to the United States and its dominant ideal of white, middle-class citizenship.

A pageant performed in early June 1917 by the public school students of Escanaba, an immigrant logging town in the Upper Peninsula of Michigan, exemplifies these tensions. Held in a lakefront park and reportedly attended by ten thousand people, the event was, according to local school superintendent F. E. King, a "cosmopolitan, broad, and tolerant" community production, researched and written by students with help from teachers and local elders. Performed as the United States was entering World War I, the pageant endeavored to "make loyal Americans who will give their lives if need be for the principles of Democracy." Presented in five episodes by students of all grades, Escanaba's pageant began with the seventeenth-century arrival of French missionaries and fur traders to the Great Lakes, represented by a group of boys rowing a boat ashore and "bearing their message of hope" to an encampment of children dressed as Indians. A "contrast dance"—a common pageant device for signaling the triumph of one civilization or social ideology over another—followed, in which a dozen "white maidens" danced opposite a group of "Indian girls" while "tepees are removed and all traces of Indian life vanish." Progressing to the nineteenth century, students dramatized the growth of the logging industry and arrival of immigrants, represented by children from local elementary schools who sang songs and paraded in the traditional folk dress of countries, including Russia, Finland, France, Scotland, Italy, Poland, and Japan. Cross-dressing and role-playing across ethnic and racial lines was part of the performance and educational experience, as members of the exclusively white student body played the roles of Indians and the Japanese, and "children of French or Swedish or Irish descent acted German, Italian or Scotch and vice versa." The parade of nations culminated, as many pageants of the era did, in an homage to "The Melting Pot," which recognized the "gifts" contributed by different nations to American society and role of the public school in "merging" these groups into a unified citizenry.

> Gifts were made by every nation
> Of the foremost traits among them.
> First there came the smiles of France,
> Then there came the Northmen's daring,
> Followed by the German's thrift;
> Then came Russia with her emblem—
> Truth of purpose—steadfast—firm;
> Ireland's trait of ready wit;
> Merged at last in this great country.
> Equal all before the law,
> Come from schools as melting pot—
> American citizens—strong and noble;
> Thus the school performs its mission.[22]

The pageant concluded with the young participants assembling on a field to form a giant star-spangled banner—a "human flag." Meanwhile, the "Spirit of Education"—an adolescent girl wearing a mortarboard and gown—sat atop a throne of flowers, flanked by peers playing the roles of Uncle Sam, who carried a sign commemorating the Northwest Ordinance of 1787, and Columbia, who carried the American flag. Anchored by these symbols of regional and national history, the Spirit of Education issued a call to the

PICTURES WERE TAKEN OF SMALL GROUPS ONLY

FROM 75 TO 100 EACH OF FIFTEEN NATIONALITIES

ALL IN HOME-MADE COSTUMES

Figure 4.1
F. E. King, *The Pageant of Escanaba and Correlated Local History* (Grand Rapids, MI, 1917).

public to defend democracy through everyday acts of good citizenship and patriotism. Reporting on the success of the pageant to the Michigan Educational Association in 1917, Superintendent King highlighted its power to "fuse racial elements . . . blending together the many warring and hostile elements into one mass, all moving impressively toward the desired consummation, a visible and tangible expression of a vital community-life."[23]

ALL NATIONALITIES BECOME AMERICAN

HUMAN FLAG

Figure 4.1 (continued)

The Escanaba pageant expressed the idea that the public school functioned as not only an institution of academic instruction but also a melting
pot—an engine of acculturation that painlessly boiled away the undesirable
features of immigrants' ethnic backgrounds while absorbing a few admira-
ble cultural traits into a "composite" American whole.[24] Popularized by a
1908 Broadway play of the same name, written by the Jewish playwright

Israel Zangwill, the melting pot metaphor offered an organizing principle for countless plays and pageants in schools, churches, YMCAs, and chautauquas in the 1910s and 1920s to assist in the mass assimilation of immigrants. It was a rite that was enacted with little subtlety a few weeks after the Escanaba pageant, on the other side of Michigan, in the booming industrial city of Detroit. There, on the Fourth of July, a group of foreign-born factory workers attended the outdoor graduation ceremony for Ford English School, a civics program established by Ford Motor Company in 1914 that was deemed so legitimate, the federal government recognized it as fulfilling some of the requirements for naturalization. Taking turns jumping into a giant model of a cauldron on a stage, the workers emerged wearing derby hats and waving American flags—a performance that symbolically (if not legally) marked their renouncement of their old-world identities and transformation into respectable American citizens.[25]

Focusing on European immigrants, and ignoring Americans of African, Hispanic, Asian, or indigenous descent, the melting pot metaphor was a powerful device for maintaining the social and political dominance of whites in the first decades of the twentieth century.[26] In addition to easing interethnic antipathies among "old stock" Americans of northern European heritage and the "new immigrants" from eastern, southern, and central Europe, it consolidated these groups—once understood as different "races" or racial subgroups—under a shared identity of white, legitimate US citizenship. As one social worker summarized the school's function as a melting pot in 1929, "[It] united the flood of ignorant immigrants and their children with their older established neighbors into an essentially solidary American citizenship."[27] This process, which scholars have described as the *consolidation of whiteness*, made Blacks, Asians, Mexicans, and other nonwhite Americans, who were migrating into cities in growing numbers after World War I, the target of a newly collectivized racial hostility.[28]

But in addition to performing this unifying and hegemonic function for whites in the domestic sphere, the myth of the melting pot shaped Americans' understandings of their emergent role on the international stage. Specifically, the narrative that the United States was defined by harmonious cooperation among people of many nations—a myth widely performed and reproduced by children in school pageants and projects—aided rising assumptions in American education, politics, and popular culture that the country was uniquely fit to model democracy and international under-standing for rest of the world.

This idea entered into the discourse of school pageants through the work of Fannie Fern Andrews, a Boston-based public schoolteacher turned social reformer who is credited with introducing the first comprehensive program of peace education into the public schools.[29] In 1908, Andrews formed the ASPL, a largely female organization of teachers, some of them affiliated with the pacifist and women's suffrage movements. The ASPL's mission was "to promote, through the schools and the educational public of America, the interests of international justice and fraternity."[30] Opposing the glorification of militarism in American schools and society before the outbreak of the First World War, the ASPL worked to transform the school curriculum from a "war orientation to a peace setting," believing that charting an alternative path of instilling "the spirit of good-will" in children at a young age was the best defense against future wars. Guided by the progressive belief that children learned best through experience, the ASPL was the first organization to offer an integrated curriculum of world-minded activities and lessons to public schools. It orchestrated nationwide student essay contests on the importance of international understanding, offered suggestions to schools for celebrating Goodwill Day (also known as Peace Day and Hague Day, held on May 18, the anniversary of the First Hague Convention of 1899), arranged letter-writing correspondences between students and teachers in different countries, and authored a four-hundred-page, yearlong curricular guide, *A Course in Citizenship* (1914), with a goal of "transforming 'citizenship education' from a national to an international perspective."[31] The book contained lessons for students in grades one through eight to develop their citizenship in "ever widening circles," beginning, as Andrews put it, with "the child acting as a member of the home, school, town or city, state, nation, and finally as a member of the world family."[32] Progressing through stories and poems about neighborliness, "the golden rule," and the contributions made by different nations to American and world history, the guide taught pupils that "even the most distant countries are closely linked to ours" and "that a citizen of the United States, the melting-pot of the nations, has peculiar obligations in strengthening the ties of human brotherhood."[33] With support from prominent benefactors such as Andrew Carnegie and Malcolm J. Forbes as well as endorsements from the NEA and US Bureau of Education, the ASPL disseminated curricular materials to thousands of schools across the country. These materials offered a blueprint for teaching an outward-looking internationalism anchored in an inward-looking celebration of American pluralism and progress.[34]

One of the ASPL's most sought-after publications was the script for *A Pageant of Peace* (1915) by Beulah Marie Dix. Andrews described it as the first peace pageant designed specifically to be performed by schoolchildren, and a participatory spectacle so vivid it would "fire every boy and girl with a permanent revulsion towards war."[35] Originally intended as a resource for celebrating Goodwill Day, the pageant quickly became a manifesto for the superiority of American democratic values on the outbreak of World War I. The pageant depicts an epic struggle between destructive forces, such as war, pestilence, famine, and crime, and progressive ones, such as peace, wisdom, justice, and social service, for influence over the human race. The production notes offer guidance to teachers on how to convey these abstract ideas to young people using the ordinary resources of school and home, including casting taller children in the roles of the evils and virtues ("naturally the superhuman beings should overtop the mortals"), and creating costumes of chain mail from dime-store pot cleaners, Grecian-style gowns from cheese cloth, and military dress from modified Boy Scout uniforms.[36] Like many peace pageants of the era, the Herculean struggle for humankind depicted in the ASPL script is ultimately resolved by the appearance of "America"—represented by a tall girl dressed as Columbia and flanked by American flags—who serves as the model for international harmony.[37] ("'How can a lot of nations all live at peace?' asks a mortal. Peace responds: 'Because they've done it.' A Teacher asks: 'When? Where?' Peace responds: 'I will show you forty-eight sovereign states that have lived in peace for fifty years.'")[38] Closing with a chorus of children singing "Battle Hymn of the Republic" and "Onward Christian Soldiers," the pageant ultimately casts the crusade for international peace in an unequivocally Anglo-American light, guiding audiences' thoughts from the evils of international war to the superiority of the American model of governance through an uplifting mix of youthful spectacle, Protestant hymns, and patriotic symbols.

Ironically, the ASPL officially abandoned its antiwar stance when President Wilson called the nation to arms in 1917, changing its name from the ASPL to the American School Citizenship League and revising its curriculum guide, *A Course in Citizenship* (1914), to *A Course in Citizenship and*

Patriotism (1918).[39] As Susan Zeiger notes, in addition to being under tremendous political pressure for conformity, Andrews's decision to support the war was ultimately consistent with her long-standing conviction that the "United States stood on the side of peace and enlightened international good will."[40] Like many peace-minded educators of her time, including Dewey, Andrews's faith in the American ideal of pluralistic democracy and belief that it held the key to world peace led her to throw her support behind the war. Active until Andrews's death in 1950, the American School Citizenship League brought a conservative vision of peace education into the public schools. While groundbreaking in its focus on demilitarizing the curriculum as well as training schoolchildren to adopt and perform a more world-minded civic attitude in school and society, it ultimately was "ambivalent about broader questions of social action and social change," Zeiger warns, and "serves as a reminder that peace education is not always or inevitably about social transformation."[41] As Andrews herself recalled of her organization's early efforts to introduce the idea of internationalism into the curriculum, "A confused sense of nationalism faced the teacher in taking up the new duty." Teaching "goodwill in home, community and national life . . . had become a recognized method of citizen training," she granted, but when attempting to teach the "beneficent effects of international goodwill, the subject suddenly became complex. It involved matters extremely delicate to handle in the classroom." Andrews recognized that "nationalism still held a commanding place in the political field," making it necessary for teachers to teach internationalism in a mitigated way that continually underscored its support of patriotism. It was necessary to "establish a balance which would give full play to the ideal yet would regulate our expectations in light of actual realities."[42]

"To Make for Better International Feeling": Interwar Performances of World Friendship and Internationalism

The two decades following World War I (1919–1939) saw a surge of organized efforts to transform American schools, and the pupils within them, into instruments for fashioning a more world-minded and tolerant citizenry. Disillusioned by the United States' failure to join the League of Nations, and the organization's inefficacy in preventing international conflict, progressives turned to the education of children as the surest route to promoting world peace. As E. Estelle Downing, chair of the Committee on International Understanding at the National Council of the Teachers of English, asked her fellow educators in 1927, "What is there to hinder us from making [children] our ambassadors to carry hope and healing to the future? We have it in our power to build in the hearts of children the foundations of a new social order, of which co-operation, harmony, and good will shall be the cornerstones."[43] Educators like Downing saw their own multicultural classrooms as the ideal "laboratory" for this social project.[44]

A common assumption in this period was that children were naturally open to cooperating with people of different backgrounds and free of the prejudices that afflicted adults. "Children know not the hateful prejudices which rankle in older minds, breeding wars and discord," wrote two school officials in *World Friendship*, a 1928 handbook on teaching international understanding in the Los Angeles public schools, one of the first public school systems to adopt such a policy. "The races must know and appreciate each other while they are young; it is too late to introduce them after they have grown up."[45] Others, including the pioneering intercultural educator DuBois, drew from a growing body of research in sociology and psychology that suggested that children were indeed prejudiced, but could be "conditioned" to adopt more tolerant attitudes through proper education. For both of these groups, instituting performances and activities in the school to teach cultural pluralism—or as the Los Angeles educators put it, "sympathetic appreciation for the accomplishments of different race groups"—offered an accessible means to engineer harmony among the heterogeneous groups of the United States and, by extension, among the nations of world.[46]

While the efforts to teach tolerance marked a departure from the rigid Americanization programs that had predominated before the war and still held currency in the 1920s, they remained deeply influenced by the worldview of the white, native-born educators who designed them. Lessons on international understanding in the 1920s and 1930s made important strides by acknowledging and appreciating cultural differences, but did not attempt to decenter the American, white, middle-class subject. Instead, they typically emphasized universalism, or the "likeness between nations"— suggesting that the world's peoples were more alike than different—and the positive "contributions" made by different racial and ethnic groups to "modern civilization," usually in the domains of art, music, science, and

invention. With a stress on uplifting the immigrant while improving relations between the United States and its trading partners and allies, these
programs often focused on understanding the peoples of Europe and, with
the development of the Good Neighbor Policy under Presidents Herbert
Hoover and Roosevelt (1928–1947), Latin America.[47]

While the migration of African Americans and Mexicans into northern
cities in the 1920s prompted a number of urban white educators to take
an interest in the "race question," most remained silent on matters of racism and segregation, and avoided drawing a connection between issues of
race and their crusade to make Americans more "world-minded." Further
muddying matters, the terms "race" and "nation" were often used vaguely
and interchangeably in educational literature for much of the 1920s and
1930s—describing, for example, Italians and the Irish as both a race and
nationality. Evolutionary and eugenicist views of race were widespread in
the 1920s, treating differences in group behavior as the result of innate biological differences and disparities in racial "fitness." This began to change
in the 1930s, however, as the expansion of the social sciences and rise of
fascist regimes in Europe led many in the American educational field to
embrace the anthropological idea of "culture" and social conditioning to
talk about differences between races and ethnicities. By the late 1930s,
when the American anthropologist Franz Boas led a national campaign
to educate the public on race and debunk Nazi racial doctrines, there had
begun a slow but significant shift in education toward rejecting hierarchical
and essentialist views of race, and teaching about diversity through the lens
of culture.[48]

The interest in promoting intercultural understanding in schools was
bolstered by the broader effort to promote international understanding
between the wars. Despite a rising tide of nationalism and nativism in the
1920s, marked by the Red Scare, resurgence of the Ku Klux Klan, and passage
of draconian immigration restrictions in 1921 and 1924, there was growing consensus within the educational field that schools needed to teach
"international good-will" and "the brotherhood of man" to prevent future
wars.[49] In 1923, the NEA convened educators from sixty countries in Oakland, California, to found the World Federation of Education Associations,
an international organization with a mission to "promote world peace
through education." The federation called on its members to reform the
educational systems of their respective countries to include "international

civics," an approach to citizenship education that would "cultivate in children attitudes of mind and habits of thought and action appropriate to effective membership in this world community." Reflecting the progressive turn toward more sensory, child-centered, and participatory teaching techniques, the organization stressed that making pupils more world-minded required not only curricular changes but also supplementing textbooks with "fresh material from other sources" and "habit-forming activities in which the children participate." To that end, one of the organization's most significant achievements was in coordinating the international observance of World Goodwill Day, building on the prewar efforts of Andrews and the ASPL. Each year on May 18, thousands of students in schools across the United States and around the world participated in peace pageants, international correspondences, essay contests, and other "ceremonies to make for better international feeling."[50]

In step with the World Federation of Education Associations, boards of education in a number of American cities and states, including Los Angeles, Rochester (New York), Kansas, Iowa, and Indiana, launched local initiatives to teach world citizenship and "international good-will" in their schools.[51] In rural and urban districts on the coasts and in the Midwest, schools partnered with local chapters of international service organizations, such as Rotary Clubs and the Junior Red Cross (JRC), and national nonprofits, like the American School Citizenship League and Committee on World Friendship among Children, to raise money for charities as well as conduct exchanges of letters, scrapbooks, dolls, and school materials with peers in other countries. Middle and high school students further developed and displayed their world citizenship by forming "cosmopolitan clubs" and "world friendship clubs"—or as one peace educator referred to them, "heterogeneous clubs"—in which they organized projects, dinners, and excursions to learn about different nations, including the "transplanted foreign life" living in their own cities.[52] Others participated in junior public forums in which they debated issues in world affairs, and the Model League of Nations, in which they studied and simulated the process of international governance.[53] According to a 1936 report in the *New York Times* by William Dow Boutwell, a staffer at the US Office of Education under Commissioner John Studebaker, these civic-minded activities sprouting up across the country aimed to "give the student a world-outlook and consciousness," and "spread the gospel of peace" in their schools and surrounding

Figure 4.2
Illustration commemorating the establishment of the World Federation of Education Associations. National Education Association, *World Conference on Education* (Oakland, CA: National Education Association, 1923), 26.

communities.[54] Years before the experience of World War II would bolster Americans' self-image as defenders of democracy and pluralism, and in a time when nativist and isolationist messages still enjoyed broad cultural resonance, these interwar school activities constituted significant strides toward developing young Americans' identity as open-minded "world

GOOD WILL to all Nations means Life, Liberty and the Pursuit of Happiness for All Mankind. It is fundamental to WORLD PEACE.

Figure 4.3
A celebration of "World Goodwill Day" at San Jose High School in San Jose, California, 1927. Swarthmore Peace Collection, http://triptych.brynmawr.edu/cdm/ref/collection/SC_Ephemera/id/%20778.

citizens" and performing this disposition as an ideal civic behavior in their communities.[55]

The shift toward teaching world-mindedness was additionally supported by the growth of the social sciences and an emerging preoccupation with the social attitudes of children. A number of influential research studies published in the late 1920s and early 1930s found that children internalized the anti-immigrant attitudes and "unintelligent nationalism" of their broader social environment, touching off concerns about their ability to develop into tolerant, democratic citizens as they matured.[56] In his research on child gangs in Chicago, sociologist Frederic Thrasher warned that the harsh Americanization of children of immigrants was driving some to delinquency, leaving them caught between the "divergent social worlds" of their parents' culture and the dominant US culture to which they were expected to conform.[57] In *Race Attitudes in Children* (1929), social worker and immigrant advocate Bruno Lasker surveyed teachers, social workers, church leaders, and parents about children's attitudes and behaviors toward people of different nationalities and races. He concluded that "race prejudice" was not instinctive in children, as was commonly believed, but rather learned through their "absorption of adult attitudes" in the home, church, school, and popular culture. He suggested that children could avoid becoming prejudiced through better education—a point that was much more readily taken up by school professionals than his accompanying recommendation: abolishing institutional policies of racial segregation.[58] In another prominent study, part of the highly publicized Payne Fund series, *Motion*

Pictures and Youth, a pair of behavioral psychologists at the University of Chicago confirmed what many educators had already suspected: children's social attitudes, including their views of foreigners, minorities, and war, were directly influenced by the movies they saw in theaters.[59] As Albert Murphy, a Protestant peace educator, summarized the problem in *Education for World-Mindedness* (1931), "The social world in which we live is teaching the child constantly—in the home, on the street; through the newspapers, the radio, the picture show; through neighborhood gossip, narrow-minded parents, and prejudiced school books. The child is not only learning from the teacher but from the world, a world which . . . awards group conformity with social approval."[60]

The emerging conclusion from this stream of research was that prejudice and social conflict might be eradicated through proper "training" in childhood, particularly through educational situations and activities designed to scientifically "counteract" the negative stereotypes and antidemocratic messages of the dominant culture. "It is as easy to train a developing child to be a citizen of the whole world of men as it is to train him to be a too prejudiced partisan of the nation on whose soil he happened to be born," wrote Helen Champlain, a psychologist, in the newly established and progressive-leaning *Parents' Magazine* in 1933. Champlain was one of many researchers to publish articles in the magazine suggesting that by exposing children, from a young age, to an international milieu of books, maps, games, pen pals, and even home decor as well as teaching them to "play peaceably with others," parents and teachers could build in their minds the foundation for a healthy, lifelong "world citizenship." As Lasker wrote in the magazine in 1928, the most powerful way to counter "obnoxious propaganda" about racial groups was not simply by exposing them to "counter-propaganda" but also to "make our children immune to all propaganda" by educating them on "moral responsibilities in human relationships . . . with methods that will render their minds flexible, agile, and capable of functioning even in a crisis." In a sign of how desirable these values were becoming in parenting circles, articles in the magazine hailed the "international-minded men and women" along with the "parents of broad outlook and international understanding" who were working to instill in their children a "recognition of the increasing interdependence of nations, an emphasis on the fundamental similarity of all peoples and a respectful interest in customs of other lands."[61]

In schools, pageants remained a popular device for conveying generalized sentiments of international understanding to students and their
surrounding communities. But some teachers worried that their abstract,
scripted, and ceremonial nature—and their associations with fervent
patriotism and Americanization—made them ill suited to dislodging the
ingrained prejudices and "national antagonisms and hatreds" that children
developed through their daily contact with peers, parents, and popular culture. A Brooklyn librarian highlighted the ambiguity of these ceremonies in
a 1925 essay criticizing the popular practice of observing "an 'international
week,' a 'peace week,' or a 'patriotic week'" in schools—noting how easily
performances of peace and internationalism slid into mindless displays of
nationalism. "The result of such whirlwind campaigns in the schools is a
grand medley of distracted teachers and bewildered pupils, getting through
somehow with their hasty, scrappy recitations and speeches," she wrote.
"No vital spark is kindled beneath the melting pot by methods of this kind,
and the melting pot does not melt until the fire is ignited beneath." What
was needed, in her view, was a "carefully studied program, extending over
months and years of our children's school life, having for its aim no narrow
Americanization of deadening uniformity or boastful so-called patriotism,
but the brotherhood of the world."[62]

International ceremonies and pageants were so common in schools that
several of the educators who responded to Lasker's survey mentioned them,
many with reservations about their effectiveness in challenging prevailing
prejudices. As one teacher reported, "An international pageant was observed
on a playground to which colored children were never admitted; but some
of these got near enough to watch it over the railing. No one in that neighborhood seemed to recognize anything incongruous about the situation."
Lasker concluded from these responses that "ceremonies and pageants to
emphasize oneness are empty gestures—and felt by children to be such—
when they stand out against systematic segregation in everyday life." He
warned further that such pageants, while performed in the name of helping
Americans of all backgrounds appreciate cultural differences, typically had
the inverse effect of propping up the "superiority" of the United States and
its native-born whites:

> Too often they engender not international friendship but unmitigated snobbery.
> "America," in the person of the oldest or prettiest or most popular pupil, drapes
> the star-spangled banner around the lesser nations of the world, as though to

protect the poor dears. Or children dressed up as American missionaries enter the jungle and cure all the ills of a Hindu community—which is represented as without the slightest vestige of civilization of its own. The Negro boy who always thought of himself as American, suddenly is forced to appear half-naked as an "African." A Chinese group, children of respectable parents, are made conscious of the failings of their nation in the matter of opium indulgence and superstition. Always, without exception almost, native American children—especially those of "Nordic" ancestry, are given an intensive feeling of the superiority of their own group over the other groups.[63]

Lasker's observations underscore how school performances promoted a vague ideal of internationalism that did not disturb white, liberal American educators' understandings of their own national and racial superiority. By avoiding discussions of racial segregation, associating "America" with whiteness, physical beauty, and popularity, assigning Black students to play uncivilized savages, and equating Asian cultures with "opium indulgence and superstition," many pageants parroted and reinforced, rather than challenged, the dominant representations and hierarchies of difference that circulated in popular culture.

In an effort to make internationalism less abstract and more relevant to students, some educators called for pageants and activities that explored the "normal life of other peoples" and Americans' "sense of debt to other cultures" in their daily lives. In a turn toward what Patricia Appelbaum calls "domestic internationalism," teachers began teaching world-mindedness in a manner that was more "concerned with everyday objects rather than with treaties."[64] They assigned students to explore their own neighborhoods and homes for signs of foreign contributions to their own lives, tracing the far-away origins of consumer goods, foods, and clothing. "All too often we teach a foreign country as if it were entirely foreign to us. We emphasize too much the differences, and too little similarities," wrote Ella Huntting, a graduate of Teachers College of Columbia University who trained geography teachers at the State Normal School of New Jersey. She called for teaching "the brotherhood of man" by sending children out to discover the foreign foods, architecture, and customs in their own locality.

In some New Jersey towns there are large colonies of foreigners living much as they did in Russia, Italy, Poland, or Czechoslovakia. If you keep your eyes open you will find Italian back yards supporting goats whose milk is used as in Italy; Greek Catholic churches with domes looking like inverted beets, much like the churches in Russia; Polish bake shops with queer looking loves of black bread,

. . . and many other interesting things. If you get the children at work to find foreigners living here as they did at home, you will find teaching European geography great fun.[65]

Other educators recommended inviting immigrant relatives and model members of the community—variously referred to as "great racial representatives" and "cultured representatives of other nations"—into the classroom to provide first-person accounts of life in their native lands. A visit from a student's German grandfather, for example, who spoke of his past as a toy maker in Germany, enchanted students and changed their negative attitudes toward this immigrant group. According to educational sociologist Joseph Roucek, a Czechoslovakian immigrant, this "drafting" of local immigrants was an effective way of teaching international understanding to native-born Americans "by utilizing our minority groups as links between America and the rest of the world."[66]

As an alternative to cultivating face-to-face encounters with foreigners, teachers could draw from a growing selection of published plays, skits, and stories to help children build "vicarious experience" of, and positive associations with, other cultures. "Small children cannot go to India and enter into the lives of the Indian rug weavers in reality, but they can get a similar result through dramatization," wrote Murphy in *Education for World-Mindedness*, a popular guide for Sunday school and public school teachers. Inspired by Lasker's research on children's racial attitudes, Murphy, who had studied under Dewey, encouraged teachers to adopt storytelling and dramatization techniques as part of a controlled "process of image-making." Here, teachers could generate positive representations and sensory experiences that would help "counteract" the negative associations with foreigners and minorities that predominated outside the school walls.

> Our attitudes-tests indicate that many people associate the word "Italian" with "garlic" and "Negro" with "inferior" and that other people associate "Italian" with "Columbus" and "Negro" with "melodies." These images are not original, they are learned somewhere by the association of the name "Italian" or "Negro" with certain pictures, actions, or attitudes. . . . If when we think of a German we see a fat and bossy individual or smell the steam of sauer-kraut, if when we think of a Chinaman we see a laundryman and smell the stifling air of a laundry, these things affect our attitude towards these races. . . . Arrange to make the first association of children with foreign peoples one of pleasure. . . . In case associations have not been pleasant, counteract these associations with many pleasant ones

and keep up the process. Emphasis upon good qualities even to the point of idealization is quite legitimate as a corrective measure.[67]

A common way in which stories and dramatizations attempted to correct white children's attitudes by showing the "good qualities" of other
peoples was by demonstrating how they materially enhanced their everyday lives.[68] In "A Boot Is a League of Nations," students learned about how a
single boot or shoe was made up of materials from multiple countries, such
as Spanish cork oak, Australian leather, and Chinese salts. Declaring "shoes
are a visible sign of international friendship," the story illustrated in simple,
child-friendly terms the liberal postwar argument for greater international
cooperation and commercial trade.[69]

A similar theme was explored in a fifteen-minute pageant, *America for
Americans*, the title of which was a critical appropriation of the nativist
motto of the resurgent Ku Klux Klan. Written by Lutheran educator Katharine Scherer Cronk, and performed in both Sunday schools and public
schools, the pageant begins with a pair of native-born American girls at
home reacting to a newspaper story about violence in an immigrant neighborhood by wishing aloud that all foreigners would be "sent straight back
to where they belong—bag and baggage."[70] Suddenly an agent appears at
the door and announces that he has come for all the foreign baggage. The
girls stand in shock as a team of agents enters the home and removes the
radio, telephone, rug, books, and china. In a moment that is revealing of
white Americans' historic tendency to simultaneously idealize and degrade
Native Americans, a pair of "Indians wearing typical costumes" then
appears and takes the lesson in immigration history to its logical end.[71]
"Red man only real American. Others all foreigners," the Native Americans
say, pointing to the two girls. They command the girls that they too, along
with the foreign household goods, must "skudoo." This idea is too much
for the girls to bear so they cry out for the foreigners to "come back, bag and
baggage." The bag agents promptly return and restore the foreign goods to
their place in the home, leading the Native Americans, resigned, to "quietly
shake their heads and leave." One of the girls calls after them in reassur-
ance, "Now don't you worry the least little bit. We're going to be perfectly
lovely foreigners after this, and really you'd rather have us than not, when
you see how much better we understand today than we did yesterday." The
other girl concludes the play by declaring joyfully, "We're all here to stay."[72]

The danger of illustrating internationalism through material goods, of course, was that it objectified other nations and cultures, erasing their humanity and rendering cultural diversity valuable only insofar as it yielded a variety of useful products to be bought and consumed. The treatment of Native Americans in Cronk's play, furthermore, highlights the consistently vague and contradictory messages that undergirded many interwar school performances about the importance of "tolerating" people of different nations and races. *America for Americans* suggested that white, native-born children should adopt no more than a generalized awareness and appreciation of both the immigrants in their midst and foreign nations from whence they emigrated. It also suggested that there was little to be done about their own group's historic and ongoing mistreatment of the original inhabitants of North America. In addition, the play proposed that understanding the United States as a "nation of immigrants," rather than acknowledging the specific histories of genocide, slavery, and discrimination experienced by indigenous peoples, Blacks (who were not mentioned in the play at all), and other nonwhite groups at the hands of whites, would be sufficient for engendering national harmony. This uplifting, broad-brush characterization of the United States as a nation built by immigrants would become even more prevalent in school pageants in late 1920s and 1930s, as educators grappled with promoting ethnic and racial cooperation to defend democracy against the rising tide of fascism in Europe.

DuBois, High School Assemblies, and *Americans All, Immigrants All*

No educator did more to harness school performances to teach about cultural differences than DuBois. Where earlier school pageants and plays emphasized the assimilation of immigrants, DuBois's productions explored how the past and present differences of racial and ethnic minorities shaped American culture while strengthening its democracy. As a white peace activist and public schoolteacher in the 1920s, DuBois became alarmed at the prevalence of nativism and xenophobia in school textbooks and society. She later became an educational sociologist, merging the science of behaviorism with school pageantry, media criticism, and other emerging educational techniques in an effort to "condition" adolescents to become more tolerant. While historians have shown how her work was foundational to the development of intercultural education, scholars of media and communication would do well to note how her work charted new ways of communicating about difference, treating scripted youth performances and cultural activities as potent, scientific instruments for "counteracting" stereotypes in a mediated society. As multiple analyses of her work have noted, her romantic view of diversity led her to develop educational materials that offered positive yet monolithic representations of cultural groups, distilling diverse cultures into one-dimensional tropes and erasing differences within them.[73] She also avoided discussing matters of segregation and institutional racism for fear of alienating the white, native-born audiences that she believed were most in need of attitude change as well as most responsible for transmitting tolerance in society at large. While she devoted her life to challenging assimilationist narratives of national conformity, the ambiguity of her materials blunted their social impact, and allowed them to be easily enlisted into projects promoting American nationalism, exceptionalism, and patriotism during World War II.

Raised as a Quaker in rural New Jersey, DuBois graduated from Bucknell University in 1914 and embarked on a career as a high school teacher. After World War I, she took leave from her teaching position to become an international peace activist, participating in the 1922 International Women's Conference in The Hague, chaired by Jane Addams, and authoring school materials about the postwar peace movement. In the same period, she traveled to South Carolina with a Quaker committee and became awakened to the plight of African Americans. Through these formative experiences, Diana Selig explains, DuBois "came to link interracial and international harmony," articulating a view that "basic to the problem of peace and war is the problem of race."[74] For the rest of her career, DuBois would dedicate herself to developing and disseminating educational techniques for reducing Americans' prejudices toward the country's "major minority groups"—including not only ethnic Europeans but also African Americans, Latin Americans, Asians, and Jews—as a necessary step toward building international peace. While she was primarily concerned with eliminating prejudices held by whites, she also saw these methods as useful for healing the "inferiority complex" held by minorities along with second- and third-generation immigrants—a problem that she and other sociologists believed was partially to blame for their "maladjustment" to society.[75]

DuBois's experiments in intercultural education began when she resumed teaching, as she started a new position at Woodbury High School in Woodbury, New Jersey, in 1924. Tasked with taking over the school's assembly program, she planned a series of assemblies to develop in the majority-white student body a "sympathetic appreciation" of racial, ethnic, and religious minorities—a format that she later called the "Woodbury Plan." DuBois also rejected the civics textbook she was assigned to teach, claiming that it was xenophobic. She lobbied to use in its place Teachers College researcher Harold Rugg's controversial social studies materials, which presented a sympathetic view of immigrant groups and, in DuBois's view, "highlighted the value of America's cultural diversity."[76] DuBois was soon to become part of a growing network of progressive thinkers and educationalists in the New York City area, including Rugg, Dewey, Lasker, Leonard Covello, and Louis Adamic, who advocated for *cultural pluralism*, an alternative to the rigid melting pot approach to Americanization that was favored by many "one hundred per centers." Rather than demand that immigrants reject their old-world identities, and conform to the "Anglo-Saxon and Puritanical impress," pluralists believed that differences should be celebrated, and that ethnic pride and American identity could be compatible with each other. Pluralism often vacillated between celebrating individual groups, on the one hand, and the exceptional nature of the United States' democratic, multicultural makeup, on the other hand. "The composite America is Scotch, Irish, English, Negro, German, Japanese, Polish, Jewish, Italian, Oriental, Russian, etc," wrote DuBois. "We have in our American life and in our public schools people from every culture in the world. We are unique in this, and hence have a unique opportunity to develop a functioning world-mindedness." Many pluralists saw the cultural gifts approach as useful for not only improving intergroup relations within the United States but laying the groundwork for peace in the international world writ large too. "Much energy has been expended in promoting the ideal of peace and international understanding," wrote Roucek. "We can simplify considerably this task for ourselves by utilizing our minority groups as links between America and the rest of the world. In other words, the best way to promote internationalism is to promote it right at home, where our idealism can be put to a practical test."[77] Ironically, the interest in celebrating the nation's multiethnic heritage may have been bolstered by the passage of the Immigration Act of 1924, which as scholars note, drastically reduced immigration, and led white progressives to take on a "romanticized vision of ethnic cultures" and "heirloom conservation approach" to teaching about their cultural differences. With new waves of immigrants effectively "locked out" of the country, more Americans became interested in preserving and showcasing, rather than assimilating away, the folkways of those who had already arrived.[78]

Still, preaching pluralism amid the anti-Communist fears and economic woes of the 1920s and 1930s remained a controversial position. While teaching at Woodbury High, DuBois continued in her peace activism, organizing Quaker conferences and developing friendships with immigrant and civil rights leaders, including W. E. B. Du Bois (no relation). Her actions drew criticism from local chapters of the Daughters of the American Revolution and American Legion, which labeled her as a "dangerous person in the community" as well as a "radical" teacher with Bolshevik leanings and antipatriotic ideas. While these criticisms only added urgency to her crusade, Selig notes that they may have also led her to adopt a more cautious approach to teaching tolerance thereafter—one that "struck a delicate balance between affirming cultural difference and assimilation."[79] By the mid-1930s, in the throes of the Depression, DuBois additionally noted the value of cultural differences as "resources" that enriched American society. "We have here representatives from almost every culture in the world," she wrote in 1935. "Should we not take advantage of these resources? Should we not urge members of each ethnic group to hold on to their differences which are socially valuable, and then to share those differences with others? . . . We are wasting our cultural resources because we, of the so-called dominant group, do not put the stamp of social approval upon those differences."[80]

At a time when nativist textbooks and ideas were commonplace in public schools, DuBois's lively school assemblies supplemented the curriculum with an unusually sustained and celebratory focus on foreigners and minority groups. Between 1926 and 1930, she authored three separate yearlong assembly programs in a series that she called "A Program for Education in Worldmindedness."[81] In the first program, "The Contributions of Various Racial Elements to Our Complex American Life," DuBois laid out the template for celebrating the "cultural gifts" of immigrants that would guide her later work in sociological research and radio. Each month over the course of the 1926–1927 school year, students researched and put on an assembly in the auditorium, described by DuBois as a "little pageant." Each assembly showcased a particular group and their "contributions" to

American society in the domains of art, music, history, food, science, and invention. The order of the assemblies roughly corresponded to relevant holidays throughout the year, with the first focusing on Italians (Columbus Day), followed by Native Americans and the British ("showing how they came together at the first Thanksgiving"), Germans (Christmastime), "Negroes" (Lincoln's birthday), Jews, and finally the "Oriental" nations of Japan, China, India, and the Philippines (in springtime, or cherry blossom season). Inspired by the ASPL's *A Course in Citizenship and Patriotism*, the Woodbury Plan aimed to build up students' world-mindedness in gradual steps and throughout the year.[82] Using what she described as a "separate groups" approach, DuBois believed that each group needed to be showcased on its own or with similar groups, and saved those that she thought experienced the most prejudice—"Negroes" and Asians—for later in the year, when students would presumably be more world-minded and receptive to their stories.[83]

The Woodbury assemblies were dynamic, participatory, and varied productions that engaged students through multiple senses. To help the student performers research stories about prominent minority figures in history and popular culture, DuBois came to school armed with "raw material" in the form of "books, magazine articles, posters, etc."[84] The students, who earned extra credit for participating, studied and performed ethnic folk dances, recited the writings of international authors and poets, prepared international foods, and played records of foreign music and languages for some nine hundred peers in the audience. The October assembly on Italians, for example, included a sketch on the influence of Latin on the English language, demonstrations of Galileo's contributions to science, and a speech, "When My Father Came to America," by a "student of Italian parentage." The April assembly on "Oriental" peoples was a panethnic performance that included a reenactment of Commodore Matthew Perry's landing in Japan, "chart talk" on Eastern imports and exports, retelling of Chinese fairy tales and the Arabian Nights, and speech on the life of Mohandas Gandhi. DuBois drew from her extensive contacts within the local immigrant and civil rights activist communities to bring in guest speakers, whom she described as "attractive leaders of various culture groups," to share stories, songs, and customs. In the February assembly on the "Negro," for instance, students in the audience were treated to spirituals sung by the school glee club, recordings of songs performed by Roland

Hayes and Marian Anderson, jokes and other examples of "Negro humor" shared by guest speaker William Pickens, a writer and field secretary for the National Association for the Advancement of Colored People, and a cooking demonstration in which a group of Woodbury girls made peanut bread with flour sent personally by George Washington Carver, then in his sixties, from the Tuskegee Institute. DuBois later wrote that the "colorful and dramatic performance" of the assembly format helped students make a stronger emotional connection to the cultures they were studying in the curriculum. They produced "a reaction they cannot gain by mere reading or by other more or less purely intellectual experiences . . . [s]ince many people in witnessing a play feel themselves for the time being to be the characters, actually living the experiences of the personalities in the programs."[85]

After chipping away at stereotypes throughout the year, the series of cultural assemblies culminated in a familiar kind of finale: a patriotic costume pageant on May 18—World Goodwill Day—titled *America and Her Immigrants*. With echoes of the Americanization ceremonies favored by assimilationists in the 1910s and 1920s, the pageant centered on the theme of the United States' history of acceptance and unity. But unlike those pageants, this one foregrounded the distinctive cultural gifts of immigrants, acknowledged African Americans (though without mentioning slavery to distinguish them from immigrant groups), and attempted to link lofty ideals of international peace to the local cultivation of tolerance and respect among groups. In it, a student representing the "Spirit of America" received another in the role of the "Spirit of Immigration," who introduced a member of "each racial element" showcased in the previous assemblies. "Each group comes before America, saying in verse form what their gifts have been, and laying at her feet, silks, laces, pottery, etc.," DuBois wrote in the script. "Some of the groups give a short folk dance before retiring." The pageant concludes with a struggle between the "Spirit of Peace" and "Spirit of War," with the Spirit of Peace emerging triumphant, followed by a group salute to the flag and lively choral rendition of "America the Beautiful." Taken together, the monthly assemblies on distinct minority groups, culminating in the patriotic finale on World Goodwill Day, reveal DuBois's romantic yet reductive view of cultural diversity as well as her effort to carefully anchor the celebration of racial, ethnic, and religious differences within accepted ideals of patriotism and national unity.

Convinced that the assembly program was having a positive effect, DuBois became interested in refining and testing the approach, and disseminating it to educators across the country.[86] In 1929, she left her teaching post and enrolled in the doctoral program in educational sociology at Teachers College. There, she became immersed in the science of behaviorism and social psychology, embracing the power of education to "condition" young people's attitudes, with teachers acting as "social engineers" of a more equitable society. She was deeply influenced by Dewey, who retired from Teachers College shortly after she enrolled, and his belief that the function of the school was to prepare students for citizenship in a pluralistic democracy. Merging these perspectives with the recent research of Lasker, whom she had personally befriended in New York City, she became convinced that students held prejudices because they were the recipients of "bad emotional training" in their daily lives, and that schools could "counteract" that training by fostering activities that appealed not only to their intellect but also their emotions and social sensibilities. "The experiences planned by us must carry an emotional tone that is strong and driving," she wrote in 1936, "because we act not according to what we know, but according to what we feel about what we know."[87] In her view, the high school assembly was the ideal "medium" for engineering such emotional experiences because it fostered positive interactions between white students and minority guests, was "colorful and dramatic" and "more emotional than intellectual," and had "controlled mass appeal."[88] Using a combination of attitude tests and anecdotal responses from students, DuBois tested a modified version of the Woodbury Plan at fifteen schools across the New York City metropolitan area and "found that the assembly afforded opportunities of giving the students the kind of vicarious experiences that tended to modify their emotional attitudes." The effect was particularly pronounced, she wrote, for "those who took part in the dramatic presentations" and acted as a member of a different cultural group: "The students who played the roles of Italian immigrants, telling why they came to America, actually lived, for a brief while, the lives of those immigrants. The Gentiles who acted in a Jewish play would never forget their experiences during the time when they were a part of the ways of thinking, feeling, and acting of that culture group."[89]

DuBois continued to test her assembly program in high schools, and published the results in sociological and educational journals. In her writings, she invited teachers to adopt not only the assembly format, which she had found to be consistently effective in producing more tolerant attitudes, but also "planned social situations" with "culture leaders" of minority groups, such as afternoon teas and home visits, to give students "intimate, face-to-face contact" with others that would create "social experiences in which the stereotypes do not fit."[90] Prejudice, she wrote, was an "attitudinal shortage" as well as a matter of "factual deficiencies" that could be corrected by bringing people into contact with individuals and ideas that challenged their biased views. She suggested that teachers, too, should meet with "cultured Jews, Chinese or Japanese, Italians or Negroes" on their own time, much as she had done in her activist work, to put them in a better "position to help counteract in the minds of their pupils the wrong stereotypes which may have been built by the church, the home, the movies, or the press."[91]

In 1934, in an effort to bring her assembly programs and educational materials to a wider audience, DuBois founded the Service Bureau for Education in Human Relations, a nonprofit agency based in a small office a few blocks from Teachers College. With a board of prominent progressive educators and string of financial backers, including the American Jewish Committee, Progressive Education Association, and WPA, the bureau operated as a national clearinghouse for disseminating DuBois's assembly programs and other antiprejudice materials to educators across the country.[92] DuBois also began teaching courses to in-service teachers in New York City, offering "techniques of social control . . . in developing more sympathetic attitudes," with a continued focus on "the outstanding cultural contribution of our major minority groups."[93] As she recalled later in life, her educational outreach took on a growing sense of urgency in this period, as "Hitler was in the background of all this."[94] Amid mounting concern that the ethnic nationalist movements in Europe could catch on in the United States, teaching cultural pluralism seemed to be increasingly vital to securing American democracy.[95]

By the mid-1930s, DuBois's ideas began to gain traction in wider educational circles. In 1934, the Commission on the Social Studies of the American Historical Association endorsed teaching "understanding and mutual toleration among the diverse races, religions, and cultural groups which compose the American nation." In 1937, the US Office of Education recommended DuBois's assembly materials in its Public Affairs Pamphlets series. Part of the federally funded Public Forum Project, the brainchild of

Commissioner Studebaker, the program organized public schools across the country to serve as venues for holding community discussions and disseminating information on current affairs, with "the educational objective of a more enlightened, better informed and more tolerant citizenry." Between 1936 and 1937, the Public Forum Project hosted over 10,000 public discussions attended by over 985,000 people.[96]

It was at this time that DuBois became increasingly concerned about the role of mass media and propaganda in shaping social attitudes, and called for schools to teach students to be more critical of the media they consumed. In an aspect of her work that has received less attention from scholars than her assemblies and other intercultural education techniques, DuBois urged teachers to help students identify as well as critique racial and cultural stereotypes in textbooks, movies, radio, and the press. After surveying textbooks in ten high schools and finding them rife with information that would produce "antagonisms" against Blacks, Jews, and southern Europeans, DuBois concluded that helping students identify the "lack of information" and "misinformation" about minorities in their media environment was as important as practicing tolerance in assemblies and social gatherings. Describing movies as "the most powerful builder of stereotypes," she suggested that teachers could engage in "positive counteraction by discussing and interpreting for the child the movies he will see in his neighborhood theatre." Children should learn to recognize how Jews were repeatedly portrayed as "crafty money-getters," Blacks as porters, entertainers, and clowns, and the Chinese as "bandits, murderers, and opium den habitués." Once trained to identify these stereotypes in the movies, students would then be able to detect them across media forms. "In respect to each group one might also analyze the stereotypes of the funny page, the vaudeville show, the play, the newspaper, etc.," she wrote. Much like her contemporary, Dale, DuBois saw stereotyping as a pervasive social problem that was exacerbated by the expansion of a profit-driven, sensationalistic, and propagandistic complex of mass communications. Both seemed resigned, however, to the idea that reforming individual social attitudes was more workable than regulating the media industries or changing social policy. "We cannot hope that this conscious or unconscious propaganda will suddenly cease," she remarked in 1935. "We can at present, only train ourselves to counteract it intelligently, consciously, and scientifically."[97]

By the late 1930s, DuBois's concern about the prevalence of prejudiced appeals in the media led her to bring her message of cultural gifts to the airwaves. She was particularly dismayed by the inflammatory radio sermons of Father Charles Coughlin, a conservative Catholic radio personality in Royal Oak, Michigan.[98] During the Great Depression, Coughlin built a devoted following of several million listeners with a stream of anti-Semitic and isolationist broadcasts, calling for a "unified Christian Front," and sympathizing with the fascist regimes of Mussolini and Hitler.[99] In 1938, DuBois approached Commissioner Studebaker about creating a national radio series to "dramatize" the contributions of immigrants to American life, much as her assemblies had done in high schools. Studebaker, who chaired the Federal Radio Education Committee and oversaw the WPA-funded Federal Radio Education Project, which promoted collaborations between educational and commercial broadcasters, agreed to the idea. The former director of the JRC and creator of the Public Forum Project in schools, Studebaker was interested in making intercultural education a centerpiece of American civics instruction and, as war loomed closer, part of the nation's core strategy of civilian defense against fascism. He had already shepherded several "educational and nationalistic programs" through the networks, including *The World Is Yours*, the natural history program on NBC, and *Brave New World*, a twenty-six-installment series on CBS about Latin America to promote Roosevelt's Good Neighbor Policy.[100] Studebaker regarded DuBois's idea for a program on immigration as a natural complement to these popular programs. In fall 1938, DuBois's organization, recently renamed the Service Bureau for Intercultural Education, worked with Studebaker's Office of Education and the Columbia Broadcasting System (CBS) in New York City to produce the series. Between November 1938 and May 1939, *Americans All, Immigrants All* aired on Sunday afternoons in twenty-six half-hour broadcasts.

As Barbara Savage has described in rich detail, DuBois found herself in frequent disagreement with the head writer appointed by CBS, Gilbert Seldes, a cultural critic and the new director of television programming at the network, over the direction of the series. Their disagreements led to a production that exhibited the tensions between pluralist and assimilationist philosophies of managing racial and ethnic diversity in the United States. Whereas DuBois wanted the series to focus on teaching tolerance and cultural appreciation to listeners, with each episode showcasing a different cultural group and targeting the "most common misconceptions held towards each specific group," Seldes believed the series should not be about "destroying prejudice" but rather offering a more unifying narrative that highlighted the cumulative processes of immigration and Americanization in US history. DuBois also urged Studebaker to hire minority advisers to the series, particularly at least "one Negro leader, not only because that is our largest minority group, but also because . . . our democracy will rise or fall according to how we treat that group." Studebaker rejected her request, but offered the prominent Black scholars W. E. B. Du Bois and Alain Locke a role as unpaid consultants. Tensions between DuBois and the production team even shaped how the series was named, with the Advisory Committee voting to change *Immigrants All*, a title that DuBois favored, to *Americans All, Immigrants All* out of concern for placing too much emphasis on immigrant identity, alienating native-born listeners, and appearing "backward looking."[101]

The episodes further reflected the compromise between DuBois's and Seldes's philosophies.[102] Half the shows were devoted to specific cultural groups, as DuBois had wished, with titles such as "The Jews," "The Slavs," "The Orientals," "The Italians," "The Negro," "The Germans," and "Our Hispanic Heritage." The other half offered more unifying, nationalistic themes, highlighting the collective contributions of various groups to American society, and in turn, how these groups benefited from the privileges of living in the United States. Among these episodes were "Contributions to Industry," which showed how "each wave of immigration contributes brain and brawn to American life"; "Contributions in Science," which explored how the United States "is at the forefront of scientific process, due to the brilliance and inventive genius of individuals of diverse racial and national origins"; and "Social Progress," which honored "champions of human freedom, drawn from many groups, [who] preserve and develop ideals for which the founding fathers fought and died."[103] Throughout the series, Savage notes, the efforts to nuance stereotypes were routinely "negated by essentialist and romantic claims" about particular groups. Some episodes, for example, made assertions about the "temperament" of some groups, such as the warmth of Mexicans, gaiety of the Irish, and financial shrewdness of Jews. But perhaps most important, Savage argues, the series' "attempts to construct a unifying theme could not overcome the reality of the historical oppression of certain groups." While episodes on Asian Americans, Mexicans, and Italians acknowledged that these groups suffered and encountered "unfriendly receptions" when they arrived to the United States, the show "presented these difficulties as the exception to the more common experience of finding welcoming shores and easy economic ascent."[104]

Despite these contradictions, the series garnered wide acclaim. The Office of Education was deluged with appreciative letters from some eighty thousand listeners, many from immigrant and minority groups that were enthusiastic about seeing their cultures represented in a positive light.[105] Overwhelmed with requests for more information, the office issued a printed study guide to refer listeners to additional readings and educational resources, including contact information for DuBois's Service Bureau as well as an illustrated map of immigrants' contributions to the country. The graphic, captioned "THEY BUILT A NATION," signaled the vast diversity of ethnicities in the United States, while also reasserting their association with a handful of "typical contributions" of material, cultural, and industrial goods, such as furniture and tulips from the Dutch, classical music from Germans; rugs from Armenians and Syrians, and folk dance and steel production from Poles and Slavs. Even more problematic, the map uncritically depicted cultural formations linked to historical policies of slavery as well as racial exploitation and exclusion, such as "Indian Reservations," Blacks harvesting "King Cotton," and "Chinese laborers" on the western railroad and Pacific coast, as part of the rosy panoply of "immigrant contributions" to the nation.[106]

Along with the booklet, schools could order scripts of the program from the Office of Education's Educational Radio Script Exchange, a lending library of popular radio scripts recently established by Studebaker, along with recordings of the broadcasts on phonograph discs, of which 1,576 copies were sold in the first year. Teachers played the recordings in their classrooms and before large assemblies, while others used the scripts to produce "mock broadcasts" of their series, with students performing the episodes in assemblies or over their schools' public address systems.[107] As a radio series loosely modeled on the high school assemblies about the "Contributions of Various Racial Elements to Our Complex American Life" that DuBois had developed a decade earlier, these school-based interpretations of *Americans All, Immigrants All* across the United States breathed new life into her vision of teaching tolerance through the participatory performances of young people.

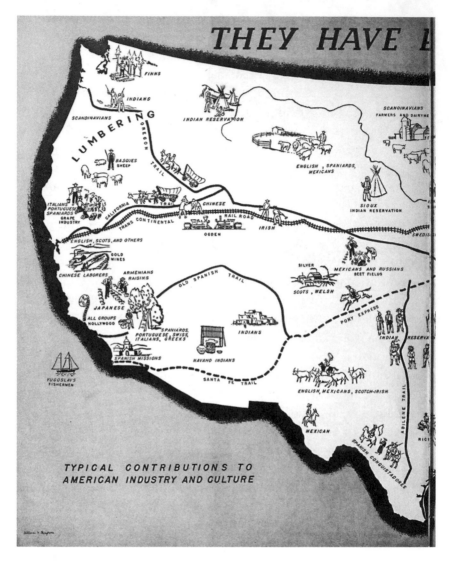

Figure 4.4
Supplementary material for the *Americans All, Immigrants All* radio broadcasts. US Office of Education, *Americans All, Immigrants All* (Washington, DC, 1939), https://archive.org/details/americansallimmi00unit.

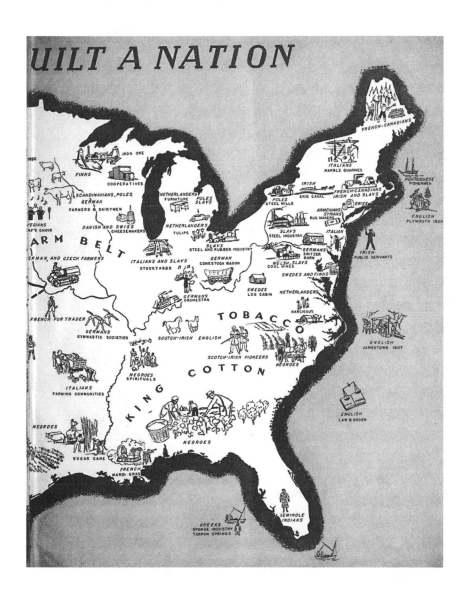

By the time the United States entered the Second World War in 1941, *Americans All, Immigrants All* was used widely in schools to teach about the "democratic way of life" and importance of defending American ideals against totalitarianism.[108] The program was accompanied by a wave of other government-sponsored radio programs, films, and teaching aids that stressed the appreciation of racial and ethnic diversity as a mind-set both necessary for victory and already essential to the American spirit.[109] In a rush to create wartime propaganda for domestic audiences, film and radio producers enlisted by the Office of War Information—such as the Italian-born Hollywood director Frank Capra, whose *Why We Fight* film series was commissioned to educate members of the US armed forces and civilians about the reasons for intervening in the war—learned that describing America's multiethnic heritage was an effective way to not only draw a contrast between democracies and dictatorships but boost morale and national unity too.[110] Such "positive propaganda" could, as one proponent put it, help Americans learn to "act as one man."[111] Wartime materials for teaching tolerance in schools also placed unprecedented emphasis on race and racial discrimination, as educators and federal agencies worked to debunk Nazi racial doctrines, and increase the participation of African Americans in the military and workforce.[112] *The Races of Mankind*, a pamphlet authored in 1943 by the anthropologists Ruth Benedict and Gene Weltfish that was subsequently taught in a number of schools as well as recommended by DuBois in her own writings, aimed to educate Americans on the science of race and dispel racial hierarchies.[113] It informed readers that race had no bearing on brain size or intelligence, and in a pointed appeal for interracial unity in wartime, that "the same blood plasma is used to restore any man of any color who has been wounded in battle."[114]

Notwithstanding this wave of wartime appeals for interethnic and interracial tolerance, though, most educational and government materials about diversity retained the prewar focus on white immigrants from Europe, and avoided discussion of civil rights, Asian exclusion, or the internment of Japanese Americans. For example, *Americans All* was often cited alongside other patriotic radio series such as *I'm an American*, which aired over NBC in 1941 in cooperation with the US Immigration and Naturalization Service, that showcased "distinguished naturalized Americans who discussed the democratic way of life." Despite claiming to be a "pageant of all America" that featured immigrants "from all lands," *I'm an American* focused on immigrants from Europe, and excluded, as Michele Hilmes writes, "those from Africa, Asia, or those already to be found upon the land before others arrived."[115] Like so many pageants of pluralism that had preceded it, the radio program reinforced the definition of the ideal, patriotic US citizen as white and of European descent.

In 1943, the NEA's *Wartime Handbook for Education* demonstrated that the ideals of teaching intercultural and interracial understanding as a basis for world peace, once seen as "radical" in the early work of DuBois, had entered the educational mainstream. The NEA stressed that schools should be teaching definitions of citizenship that positioned intercultural understanding as a bedrock of modern American democracy. Declaring that "the day of isolation has passed," the organization urged,

> Study of United States history should include study of our nation's world relations; world history should include study of international organization and cooperation; elementary-school courses should include sympathetic study of other peoples, with increased attention especially for Canada, Latin America, India, China, Russia, and the British Commonwealth of Nations; imperialism and colonialism should be re-examined. . . . Racial and national prejudice and misunderstanding must be reduced. Contributions of other races and nations should be studied; fallacies of Nazi racial theories should be exposed; the study of ethics and the moralities of great religions should be encouraged; restrictions on school study of the languages and cultures of the people with whom we are at war should be discouraged.

In addition to positioning intercultural understanding as a prerequisite for victory in war and postwar peace, the handbook stressed the role that youths should play in transmitting these ideals, noting that "assemblies and student organizations must aid in civic education" in the school and community. Schools should devote at least two hours a week to "student forums, assembly programs, and club discussions . . . as a means for clarifying the issues and conduct of the war and of plans for peace," the handbook stated. "Pupil leaders and organizations should be given responsibility in the development of morale and enlightened opinion in the community. The importance of music, art, and ceremonial in strengthening emotional support for the civic drives of pupils should be recognized."[116] In her own wartime writings, DuBois continued to tout the cultural gifts approach as a means of building tolerance and strengthening democracy, arguing that the school was the "most logical medium for reaching the children and youth of plastic age," and could "translate abstract concepts of democracy into concrete realities in the daily experience of American children, and thus instill positive attitudes toward democracy."[117]

DuBois's work highlights how a growing progressivist concern for teaching "world-mindedness" before World War II led to the configuration of schools and students as critical instruments of global as well as intercultural information and ideology in American culture. Spurred by concerns that the speeding up of cross-cultural contact, international war, and mass communications in the new century was exacerbating prejudices and straining the social fabric, educators turned to not only the new mechanical media technologies but also humble "devices" like homegrown pageantry and youth performances to produce their own visions of a democratic, harmonious nation and world. Through intercultural peace pageants, skits, and school assemblies, educators found an accessible, malleable, and participatory medium for transforming the attitudes of youths and reforming society at large. They saw these performances as not only a way to teach lessons in good citizenship to children of diverse backgrounds but also to directly "counteract" the informational inadequacies of modern society and mass culture, providing, as DuBois wrote, "those elements of culture that are not being successfully transmitted by other agencies."[118]

In these ways, children's internationalist performances in school were an important educational technology, serving to instruct both the children and the communities for whom they performed. But like most technologies, these devices were shaped by the values and worldviews of the people who designed as well as used them. Presenting a vision of democratic multiculturalism that directly descended from the celebratory myth of the American melting pot, these performances showcased a limited range of cultural differences while centering Anglo-American identities. In addition to reducing complex cultures into a handful of traits easily graspable and performable by children, this approach presented a utopian vision of harmonious cooperation that actively obscured the ways in which minority groups faced ongoing discrimination and exclusion in the American school system and society. In these ways, youth pageants and performances of peace reproduced many of the problems of the dominant culture that they endeavored to reform.

5   Pen Pals and Messenger Dolls: The Rise of Mediated Intercultural Exchange

I thought that foreign children
Lived far across the sea,
Until I got a letter
From a boy in Italy.
"Dear little foreign friend," it said
As plainly as could be.
Now I wonder which is "foreign,"
That other boy or me?

—Ethel Blair Jordan

In spring 1927, an American teacher named Lillie Newton Douglas traveled from Los Angeles to Tokyo to initiate an exchange of visual materials between Californian and Japanese schools. Sent by the Visual Education Department of the Los Angeles school system, which was at the forefront of the movement to teach international understanding to young people, Douglas's mission was to make arrangements with the Japanese to exchange "student-made school work, handicrafts, drawings and objects" to illustrate the "home life, school life, and play life of the two countries."[1] Conducting such an exchange with Japan was important, educators believed, to counter the anti-Asian rhetoric that had led to the passage of the 1924 Immigration Act, which barred Asians from immigrating to the United States. Yet while she was in Japan, Douglas encountered a similar program that was already underway on the ground—involving not an exchange of visual aids but rather a shipment of thirteen thousand dolls. Dubbed "messengers of goodwill," the dolls had been outfitted by thousands of American schools, scout troops, and churches, and sent by steamship to Japan under the auspices of the Committee on World Friendship among Children (CWFC), a non-profit peace education organization based in New York City and supported by the Federal Council of Churches, an ecumenical association of Protestant denominations. Fortunately for Douglas, her educational credentials earned her an invitation to an official reception for the dolls held aboard the steamship *Tenyo Maru*. While attending the reception, she was introduced to Japan's vice minister of education and seized the opportunity to make her pitch for an exchange of visual aids with the Los Angeles schools. He was impressed by her plan, and within a few days, ten Tokyo schools signed on. Later that year, Japan would send not only its own shipment of *Torei Ningyo* (dolls of gratitude) back to the CWFC in New York but also several large crates of visual materials, including games, picture books, clothing, and children's art to the schools in Los Angeles.[2]

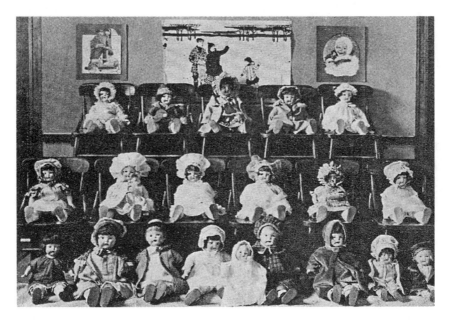

Figure 5.1
American dolls await shipment to the children of Japan in 1927 as part of the Friendship Doll Project of the Committee on World Friendship among Children. Sidney L. Gulick, *Dolls of Friendship: The Story of a Goodwill Project between the Children of America and Japan* (New York: Committee on World Friendship among Children, 1929), 10.

Though originating on opposite coasts of the United States, the concurrent efforts to exchange dolls and assorted visual media between American and Japanese children in 1927 were outgrowths of the same idealistic vision that emerged in American education in the first decades of the twentieth century. As educators strove to incorporate new instructional media and global perspectives into their teaching, many embraced long-distance exchanges of media and materials among children as a way to expose Americans to different cultures, and challenge dominant prejudices and stereotypes. Long before the arrival of computers and the internet would, at the end of the century, usher in an explosion of global "virtual exchange" practices in schools, American schoolchildren swapped tens of thousands of messages with peers abroad through media as varied as letters, scrapbooks, radio greetings, and token objects such as dolls, toys, and seeds. What united these diverse circulations of media and materials was their perceived ability to enhance school lessons about the nation and international world, promote progressive pedagogical objectives of civic-minded, hands-on learning, and convey sentiments of peace and amity between citizens of different nations, cultures, and religions.

These youthful, grassroots exchanges were seen as such an effective instrument for shaping international attitudes that educators and policy makers began promoting them as a form of state-sanctioned junior diplomacy after the First World War. The growth of international exchanges in schools before World War II thus laid the conceptual groundwork for later federally supported initiatives to emerge, such as the People-to-People Program, a Cold War citizen diplomacy effort established by the Eisenhower administration in 1956, and more recently, Exchange 2.0 and the Stevens Initiative, developed by the United States Institute of Peace and the US State Department under the Obama administration, respectively, to prevent conflict and combat global extremism through youth dialogue over the web.[3] Together, these initiatives point to a long history of Americans approaching citizen-based communication across borders, through diverse visual and textual media, as a tool for not only teaching tolerance and world-minded citizenship but wielding national influence and soft power too.

This chapter examines these contradictions in four practices of international exchange among children that emerged from the turn of the century through the Second World War. These include the international correspondence movement and early pen pal programs, exchange of children's "goodwill greetings" over the radio in the 1920s and 1930s, interschool exchanges of student-made scrapbooks through the JRC, and exchange of dolls and other "tokens of friendship" through the CWFC. I argue that these heterogeneous textual practices together constituted an emergent and influential form of global participatory communication that I call *mediated intercultural exchange*. Mediated intercultural exchange refers to the organized transfer of text, images, sounds, and/or material objects between citizens of different nations and cultures for the purpose of bridging geographic, cultural, and linguistic barriers, exchanging cultural knowledge and messages of goodwill, and communicating cultural and national values and ideologies. The ways in which American students engaged in this practice in the early twentieth century spanned multiple networks and modalities of communication, including letters sent through the postal system, speeches and songs broadcast over the radio, and material goods shipped overland and by sea.[4] Such exchanges laid the foundation for practices of virtual intercultural exchange that continue to thrive in education today, primarily in digital rather than analog forms.[5]

In ways that resonate strikingly with discourses about the digital age, the first decades of the twentieth century were hailed by social observers as a new era defined by rapid technological advancement, ubiquitous communication, and global connectivity—an "age of super-contact" and a "shrinking world."[6] Commentators variously celebrated and fretted over the global proliferation of new technologies that enabled ordinary citizens to both *consume* unprecedented amounts of information about the world—whether by newspaper, newsreel, movies, or radio—and directly *contact* as well as communicate with distant others.[7] In this climate, international correspondence among youths emerged as both a popular pedagogical device and reassuring symbol of the utopian potentials of a world rendered "smaller" by technological modernity. As Florence Brewer Boeckel, the director of education for the National Council for the Prevention of War, described it in 1928,

> With the coming of steamboats and locomotives fast mails were built up, and now, by telegraph, cable, and by radio it is possible to send a message all the way around the earth in a few moments. It is not only grown people who send messages back and forth on business, but from children too letters are being sent in all directions, asking questions about what it is like to live in other countries.[8]

In contrast to the divisive and aggressive rhetoric of geopolitics, and the sensationalist content of movies and radio, educators regarded international exchanges among children as a wholesome, innocent, edifying, and democratic activity that would help bring up a well-rounded, world-minded citizenry and more peaceful future. As a high school principal in Kansas observed with pride in 1936:

> One can scarcely question the evidence of a developing international mindedness and friendship, when one sees in his own school a pupil happy in the task of developing a music portfolio for a pupil in Denmark, another helping a friend in South Africa to enlarge a stamp collection, another making a scrap book for a friend in the Philippines, while another writes with passion in his soul on the subject of World Peace and the prevention of war.[9]

Yet while conducted in the name of progressive education and international peace building, these exchange practices served not only the pedagogical aims of educators but also the interests of technology industries and governments. Recognizing the power of children's voices to spread information and ideas across borders, policy makers and promoters of new technologies began utilizing goodwill exchanges as a way to promote American values and technological products. For example, early pen pal programs and the scrapbook exchanges of the JRC were praised by the US Office of Education for aiding America's mission to "make the world safe for democracy" after World War I. Additionally, staged exchanges of "goodwill greetings," transmitted by children in different countries over the radio and by telephone in the 1920s and 1930s, were orchestrated by radio networks and telephone companies to demonstrate how new communications technologies could contribute to world peace. These state- and industry-supported efforts to foster mediated intercultural exchanges among youths have become commonplace in recent decades. From email exchanges conducted between schools in the United States and Union of Soviet Socialist Republics to thaw tensions during and after the Cold War, to more recent efforts to "wire" America's schools and "connect the classroom to the world" through a suite of digital technologies, mediated intercultural exchanges and the equipment that facilitate them have been increasingly touted as necessary components of an education in a globalized and networked society.[10] All this is to say that mediated intercultural exchanges among children have been championed not only for educational but also industrial and political purposes for over a century. They have, in many instances, harnessed altruistic signifiers and grassroots practices of youth communication, intercultural curiosity, and feelings of goodwill into the service of shoring up commercial technologies and state policies.

International Correspondence as Junior Diplomacy

Modern-day exchanges of media among schoolchildren have roots in the international correspondence movement that emerged in Europe and the United States at the turn of the twentieth century.[11] Led by teachers and professors of modern languages and geography, the goal of the movement was to encourage adolescents and university students in different countries to write letters to each other to improve their foreign language skills and knowledge of each other's countries.[12] In the 1890s, teachers in secondary schools, lyceums, and universities formed "international correspondence bureaus" to match interested students with peers in England, France, and Germany. The success of the movement in Europe inspired faculty in modern languages at Swarthmore College in Pennsylvania, with support from the Modern Language Association of America, to create an American bureau for matching high school and college pupils in the United States and Europe in 1899. In 1901, the bureau at Swarthmore received 590 applications from American students seeking language partners in French, German, Spanish, and Italian.[13]

From the beginning, proponents of international correspondence argued that it enlivened schoolwork and promoted an unprecedented feeling of "kinship" among students of distant nations. "One may almost say that it gives the work a new meaning, changing it from the dull study of a dead unspoken language to that which may be called a living language indeed," wrote Edward Magill, a professor at Swarthmore who chaired the International Correspondence Committee of the Modern Language Association. He noted how young people's participation in international correspondence could enhance not only their academic knowledge but their long-term development into world-minded, prosperous citizens too. "It goes on from year to year, long after leaving school or college, and may lead to many new friendships and business relations, and thus be a life long [sic] source of pleasure and profit to those thus engaged."[14]

In a manner that parallels today's discourses about the internet, proponents of international correspondence celebrated both the technological modernity and grassroots authenticity of this emerging communication practice. Through "this friendly intercourse," wrote Josephine Doniat, a high school teacher in Illinois, in 1904, "the student gains a fund of information about the manners and customs of the foreign nation, and as he grows to respect and even love his partner, he realizes that foreign modes of life, while they may be different, are not, on that account, inferior to his own." She described the rise of international correspondence among youths as a sign of the era's rapid advancements in science, technology, and communication, themselves a product of "this age of scientific research and scientific results, when people at opposite ends of the globe are in closer communication with one another than was once possible between the inhabitants of neighboring towns."[15] At the same time, advocates praised the humble and personalized nature of letter writing, portraying it as uniquely effective in transcending linguistic and cultural barriers. International correspondence revealed not only the inherent sameness of people around the world but also their universal desire to learn about one other. As one German schoolmaster observed,

> It shows the pupils that beyond the frontier, "beyond the mountains," there are boys and girls living, who are toiling like them in order to learn the foreign language, who meet with the same difficulties in overcoming the obstacles which they find in their way. . . . Since so many thousands are corresponding with each other, we may say that the different countries which, up to this date, were used to consider each other as mortal enemies, only thirsting for each other's blood, . . . are linked together by new ideas, by peaceful thoughts, by mutual respect and love.[16]

The personal touch of letters, it was believed, could imbue traditionally "dry" school subjects like languages and geography with "the pulsation of life," making it possible for students to form friendships that could engender lasting feelings of sympathy for their brethren overseas.[17] Still, despite the praise for its global reach and democratic nature, the movement was largely confined to western Europe and the United States, and often subjected participants to strict academic guidelines and adult oversight.

American participation in the international correspondence movement remained relatively low until after the First World War, when teaching internationalism became a new priority for progressive educational reformers.

Additionally, the war had elevated Americans' interest in international correspondence through their churches, schools, and voluntary organizations, as many of these sent relief supplies to Europe and received letters of thanks in return. Some of these shipments of aid resulted in friendly correspondences that lasted for years.[18]

The American role in the movement grew sharply with the establishment of the National Bureau of Educational Correspondence at the George Peabody College for Teachers in Nashville, Tennessee, in 1919. Founded as a postwar peace initiative with funding from the George Peabody Foundation, what became known as the "Peabody Bureau" was the first large-scale program to promote international correspondence among children in the United States and abroad.[19] Its original goal was to coordinate exchanges of letters between students in the United States and France to commemorate the wartime alliance between the two nations. Within the first year, the bureau matched an estimated eighteen thousand pairs of students, taking extra pains to ensure that there would be an even geographic distribution of participants so that a single schoolroom of French letter writers would receive letters from various parts of the United States and vice versa.[20] The program grew quickly with endorsements from the US Bureau of Education, NEA, National Council of Jewish Women, International Association of Rotary Clubs, and General Federation of Women's Clubs. It was particularly active in the American Midwest and parts of the South, establishing local chapters in Iowa, Missouri, Oklahoma, Illinois, Indiana, Minnesota, Kentucky, Tennessee, Mississippi, Kansas, and Texas. By the mid-1920s, the Peabody Bureau arranged thousands of matches between American children and peers in not only France but also China, Japan, Germany, Spain, and parts of Latin America.[21]

Promotional literature for the program emphasized its importance as both an edifying social practice for children and promising new form of citizen diplomacy. Receiving a letter from abroad, written in French and marked with a foreign postage stamp, was the first opportunity most American children had to see "living French with a personal throb in it," wrote the program's first director, A. I. Roehm. Whereas earlier international correspondence programs had emphasized the exchange of proper, staid letters filled with academic information, the Peabody program prided itself on fostering a more modern approach to correspondence that allowed for "personal touches," including information about students' individual interests and engagements with popular culture. Children were encouraged to enclose with their letters an array of other visual and textual materials such as postcards, photographs, pressed flowers, songs and poems, newspaper clippings, and "specimens of handiwork."[22] This "wealth of enclosures," wrote US commissioner of education Philander P. Claxton in his endorsement of the program, would ensure that correspondents gained not only academic knowledge of the country but also an enriched appreciation of its people and culture. "Along with the letters," he noted in *School Life* magazine in 1919, "there will be a fine exchange of historical, artistic, geographical, manufactural, commercial, and home-life material and information, clippings, picture postals, kodak [*sic*] views, etc., leading up to the deepest exchanges of human sympathies and ideals, that will reinforce international good will."[23]

But the newfound enthusiasm for fostering individualized cultural exchanges among youths was tempered with constant reminders about the seriousness of international correspondence as part of a young person's citizenship education and extension of the high-level diplomatic workings of the state. To advance the program's mission of promoting peace and preventing future wars, American and French officials maintained that proper representation of one's self and nation was of utmost importance. Children in France were instructed to behave as "cultured, careful and correct English letter-writers."[24] Additionally, boys were matched exclusively with boys, and girls with girls, to prevent improper romantic ties from forming.[25] In apparent contradiction of other official accounts encouraging children to enhance their letters with personal material, one program official instructed American children to avoid using an overly informal tone and focus instead on exchanging "real information about the foreign country . . . not merely the description of everyday happenings."[26] As Charles Garnier, the inspector general of the French Ministry of Public Instruction and head administrator of the Peabody program in France, put it in 1929, educational correspondence was "not simply fun and play":

> It must not, cannot be an everyday, irrelevant, hasard [*sic*] exchange like "phoning to the next village neighbor." . . . When we correspond with another country, we must always write what we know and what we think worth sending across the ocean, worth being told to the son of another republic and it must be what is best in our beliefs and thoughts. Why? Because, do what we may, when we write to the member of another school abroad, we do write a little, though informally, as a

member of our own school; when we have the honor to write to the citizen, however young, of another State, we do write a little as a citizen of our own state.[27]

Garnier was one of many government and educational officials to cast
international correspondence as an exercise of citizenship development
and junior diplomacy, noting that the letters and materials mailed among
young people could be as influential to the future of international relations
as the official dialogue between political leaders.[28] He encouraged the young
American participants to imagine their communications as akin to those of
US secretary of state Frank B. Kellogg, who at the time was soon to arrive to
France to sign the Kellogg-Briand Pact renouncing war and promoting the
peaceful settlement of international disputes. Although students wrote with
a "humble steel pen or fountain pen," he mused, if they behaved as respectable representatives of their nation, their pens would "at once, in your very
fingers, be changed into gold and be worth being put side by side with Kellogg's pen as helping to spread good will and peace."[29] Such exhortations
from both sides of the Atlantic testified to the view that the communiqués
of children, though valuable precisely because of their youthful authenticity, could nevertheless influence nations' perceptions of each other, and
should therefore be composed with great care and adult guidance.

Praised as vivid supplements to textbooks and other visual aids, the
letters and diverse enclosures exchanged by schoolchildren served as
important global texts in the classroom. Missives from abroad were enthusiastically scrutinized and studied by students. Teachers displayed them on
bulletin boards and in hallway showcases, or filed them away for future
use as authoritative reference materials on foreign languages and customs.
Representatives of the Peabody Bureau cited the power of exchanged materials of all types—even the handwriting and exotic postage stamps on the
envelopes—to "vitalize interest" in people and places abroad.[30] The act
of creating, collecting, and studying exchanged materials simultaneously
addressed two objectives of progressive education: the development of
world-minded citizenship, or an understanding of and appreciation for the
differences as well as interdependencies among nations, and the promotion
of more active, child-centered, and multisensory forms of learning.

By the late 1930s, young Americans could find pen pals in and outside school through a growing number of nonprofit organizations. These
included the American School Citizenship League, YMCA, Pan-American

Union, JRC, and Children's Caravan, an international correspondence club founded in New York City by the Baha'i leader Mirza Ahmad Sohrab.[31] Children could also find and request international pen pals in a number of newspapers and magazines, including *Boys' Life*, the official magazine of the Boy Scouts of America; the *Rotarian*, the publication of the International Association of Rotary Clubs; *Everyland*, a children's magazine produced by the Central Committee on the United Study of Foreign Missions; the *Christian Science Monitor*; and the *Cleveland Press*.[32] As the *Chicago Defender*, the nation's leading Black newspaper, grew in circulation across the United States, Caribbean, and Africa, it established chapters of a youth organization called the Billiken Club that performed the rare service of connecting African American children with writing partners in other states and countries. Publishing young readers' pleas for new pen pals and their reports on successful correspondences in the junior section of the newspaper, the *Defender* highlights how Black children were largely excluded from white-led initiatives to promote world citizenship, and given separate and fewer routes to developing international connections than their white peers.[33]

As correspondence grew in popularity in the 1930s, the terms "pen pal" and "pen friend" gradually began to replace the formal moniker of "international correspondence" that had defined the movement since the beginning of the century. Whereas international correspondence emphasized restrained, academic, and often adult-supervised discourse among school pupils, the informal and child-oriented terminology of pen pal programs reflected a conscious turn toward elevating young people's individual voices, interests, and everyday experiences in the promotion of international understanding. For some advocates, it was precisely the individualized nature of pen friendship that made it so effective in communicating the American ideals of democracy and freedom of speech to people in other parts of the world. In 1930 in Boston, Edna MacDonough, a student at Wellesley College, founded the International Friendship League, a pen pal matching agency that coordinated the exchange of a reported *five million* letters between young people in eighty-six countries within its first decade.[34] Touting the league's correspondences as "entirely unsupervised and . . . therefore unstereotyped," MacDonough believed that fostering candid exchanges could help people of different nations supply "true pictures" of each other, and in this way, overcome the "false prejudices" propagated by textbooks, movies, and the press:

> The League feels that the young people, when they have grown up and taken the reins of the government in their own hands, will not have merely abstract ideas about other countries which they have gathered from school books or newspapers, but their ideas will be vitalized by personal experience. . . . The young people write about their schools, their homes, their churches, athletics, vacation; in fact, they give true pictures of their lives in general. In many cases, pictures and souvenirs are exchanged.

Pen pal exchanges would help other nations "learn the truth about the fine ideals and standards of our great American democratic civilization," she wrote, while correcting Americans' "erroneous impressions of our foreign neighbors" such as "Mexico as a land of barbarous bandits, the wonderful Hawaiian as a ridiculous hula girl, and China as a land of the ignorant superstitious coolies."[35]

Solidifying the turn toward informal, child-driven pen pal exchanges in schools, in 1942 the Peabody Bureau departed from its earlier guidelines of emphasizing proper and restrained academic discourse between pupils. Instead, it instructed teachers that exchanging "episodes of home life and school life, expression of likes and dislikes," and "real sharing of joys and sorrows" between students was "how it should be, for personal attachment . . . must never be crowded out of the correspondence, no matter what we educators may wish to outline for these young people in the way of topical information on social conditions, in order to make the correspondence 'educational.'"[36]

In the buildup to World War II, educators began to tout pen pal correspondence as a strategic tool for awakening American children's interest in global affairs, boosting their literacy of world news and maps, and humanizing the costs of international conflict. In 1939, Mary MacDonald, a high school teacher in Monson, Massachusetts, credited her students' European pen pals for helping them become more interested in and informed about the crisis brewing overseas: "They are getting the point of view of children who are in the heart of world-troubles and yet have wishes and ambitions like their own. They are hunting in the newspapers for items of those countries where their 'pen pals' live. They give oral topics on the customs they have learned. Estonia is now a real place on the map. Latvia is more than a name."[37] When the United States entered the war, the US Office of

Education and US Department of State urged educators to prioritize student letter exchanges with America's allies, particularly the British Empire and "other American Republics" of Latin America, to strengthen mutual understanding and morale.[38] Teachers could find exchange partners for their students by writing to the Department of State, Division of Inter-American Educational Relations at the Office of Education, and private organizations such as the English Speaking Union and Pan-American Union.

School officials also underscored that writing to a pen pal could boost critical skills in media and information gathering, appraisal, and communication in wartime. "Softpedal the old-type, formal, textbook question-and-answer technique, and substitute thought-provoking, purposeful pupil activity in its place," wrote an administrator in a wartime guidebook to social studies teachers in Stanislaus, California, in 1943. Writing to a pen pal, he asserted, along with related "live" activities, such as scrapbooking news articles, creating bulletin boards, studying maps, and preparing talks on global affairs, could transform the social studies of wartime into something "vital, real, and stimulating" for students. Furthermore, two educators in Ithaca, New York, maintained that pen pal correspondence could build a "wholesome nationalistic attitude" in American youths, encouraging them to reflect on their nation's unique history of social progress, its democratic system of government, and "our interdependent relations with the people of other nations" as a vivid contrast to isolationism and totalitarianism.[39] As testaments to children's diverse individual interests and international relationships, pen pal exchanges came to represent, for advocates in the United States, the openness of American society and its respected position as the torchbearer of liberal democracy. These ideas would later come to shape American foreign policy after the war, as the Truman and Eisenhower administrations officially embraced cultural exchange and "cultural relations" as vital to national security interests as well as the expansion of the free world.[40]

Children's Radio Greetings and Technological Goodwill

The rising popularity of international correspondence programs in the interwar years catalyzed a number of experiments in mediated intercultural exchange beyond the postal system. One variant took place over the new medium of radio, in the form of "wireless messages of friendship" recited annually by children in different countries on World Goodwill Day.[41] The tradition began in 1922, with a brief radio address delivered by a group of children in Wales, under the direction of the Welsh pacifist Reverend Gwilym Davies. In the following years, short declarations of peace and international goodwill were transmitted each May by delegations of young people in multiple countries, including Sweden, Japan, Poland, and the United States. In the 1930s, NBC carried the broadcasts into American schools and homes as part of its "World Goodwill Day Program." In their speeches, the young speakers called for "a better and more peaceful world," celebrated science, technology, and the League of Nations, and in the case of the Welsh speakers, decried war as a folly created by adults misdirected by their own nationalism.[42]

In 1937, a more elaborate and musical series of "good-will greetings" was transmitted to over a million children across the United States by the American School of the Air, an educational radio program produced by CBS under the directorship of Helen Johnson. CBS's César Saerchinger, a musicologist and the first American radio reporter stationed by a major network in Europe, visited twelve countries to produce the series, which he called the "first series of transatlantic radio broadcasts ever designed for schools." From radio stations in London, Paris, Turin, Budapest, Berlin, Warsaw, and Prague, youth choirs and orchestras performed folk songs and spoke greetings over the ether to their American brethren overseas. The broadcasts reportedly transmitted a "greeting so real, so close," that some American students, listening in their classrooms, shouted greetings toward the radio sets and proceeded to write "a flurry of letters in childish script" to send to their European peers in the mail.

While the Welsh wireless messages were explicitly antiwar, the CBS broadcasts were decidedly apolitical and avoided any reference to the rise of aggressive fascist regimes in Europe. Saerchinger later recalled that he had to prod the German choir, led by a "brown-shirted Nazi," to sing traditional German folk songs instead of the "new Nazi marching songs" that they preferred.[43] The result, the *Atlanta Constitution* reported approvingly, was broadcasts that contained "never a hint of propaganda. To keep that out in these tangled times is a lesson in diplomacy itself. . . . [E]very one of the broadcasts has been relieved of the taint of politics—merely a charming interchange of the things that children like and understand."[44] The goodwill broadcasts thus emblematized a core contradiction of mediated intercultural exchanges that was central to their popularity in the twentieth century. Regarded as innocent communications among children, proponents often described such exchanges as simultaneously contributing to the resolution of geopolitical problems while remaining removed from politics and above the political fray.[45]

Instructing children to sing or speak messages of goodwill over the radio was not an entirely new phenomenon. It was part of a long-developing American tradition of public demonstrations of new electric technologies that endeavored to convince the public of the social benefits of a more wired and networked world. From demonstrations of the power of telepresence via telegraph and telephone stunts at turn-of-the-century world's fairs, to the International Radio Broadcast tests of 1923–1926, an effort, coordinated by the Radio Corporation of America, General Electric, and other leading players in the radio industry, to mobilize radio listeners to receive international broadcasts at multiple locations across the Western Hemisphere and Europe, promoters of new technologies continually sought opportunities to publicize the border-crossing and peace-building power of their innovations.[46]

But the interwar years saw a conspicuous growth in the number of technological demonstrations performed by children. In addition to the goodwill radio broadcasts, a "good-will chain" of greetings was relayed by children in Washington, DC, to peers around the world via telephone on World Goodwill Day in 1931. This "around-the-world telephone conversation," coordinated jointly by the World Federation of Education Associations, National Council for the Prevention of War, and telephone companies, drew a direct link between the educational ideal of fostering international understanding among youths and the global expansion of Western telecommunications systems. The ceremony was a response to "the great problem of our times," remarked Augustus O. Thomas, president of the World Federation of Education Associations. That was "to realise the new kind of world, closely united and interdependent, capable of quick communication for the adjustment of any differences or misunderstanding in which we live. It is believed that to talk in this way around the world, to hear each other's voices across thousands of miles, will help young people in the schools to form a new picture of the world."[47]

Events like these can be characterized as expressions of an emerging ideal of *technological goodwill*—an optimistic belief that new, networked communication technologies, when placed in the hands of citizens, and particularly children, would be a force for building international understanding and eradicating prejudice through increased human "connection." These events signaled the increasing coordination between educators, industries, and government agencies to harness technology for civic uplift and public diplomacy—applications that helped to further legitimize the technologies themselves. In 1932, as the children of Wales proclaimed in their eleventh radio address, the "miracle" of radio made the world feel smaller and more intimately connected—"like a big village" and "world neighbourhood." They urged their peers to build a future in which radio messages would flow freely and communication across borders would forever be friendly. "Now the air carries music from many lands and voices in every language and through our radio services nations may be closer friends," they said. "Let us, then, boys and girls, in thought, word and deed, strive with all our might that the messages sent from our own countries shall always be messages of friendliness and of goodwill."[48]

The US Office of Education, promoting NBC's World Goodwill Day Program to schoolteachers, echoed the idea that these high-tech exchanges between children could mold them into world-minded citizens and usher in a future of peace. Lauding "this friendly, neighborly deed, made possible by science, which like childhood, knows no national boundaries," an official declared that children would "be led to feel the thrill of companionship and fellowship with other little folk the world over and that will surely have some influence on their actions when they have grown to be men and women."[49] The broadcasts thus symbolized the faith that American educational leaders were beginning to place in young people's use of new technology to transform the social order. Like the emerging technology of radio, children were full of potentiality, and in the minds of peace advocates, could become powerful instruments of social reform. With proper education and access to new tools of global communication, children could cultivate feelings of friendship and goodwill that transcended borders—and compensated for the failings of governments, diplomats, and the grown-up world.

JRC Albums and American Humanitarianism

In addition to exchanging letters through the mail and greetings over the radio, tens of thousands of American students exchanged handmade scrapbooks through the international humanitarian organization, the JRC. Founded in Canada in 1914 to mobilize youths to assist the Canadian Red Cross in relief efforts during the war, the JRC soon established chapters in the United States when the country entered into the conflict in 1917.[50] It quickly became the largest nongovernmental organization to have a presence in American public schools. At the beginning of the school year in 1917, Woodrow Wilson, who as president of the United States also held the position of president of the American Red Cross, issued an invitation to the schoolchildren of America to join the JRC, noting that it would bring "opportunities of service to your community and to other communities all over the world." From its origins in wartime, the organization was regarded as an institution that taught good citizenship to children by providing opportunities for "service for humanity" and enacting the progressive adage of "learning by doing." By gathering relief supplies for wounded soldiers and displaced civilians, Wilson told the children that "more perfectly than through any of your other school lessons, you will learn by doing these kinds of things . . . to be the future good citizens of this great country which we all love."[51]

During the war, between eleven and thirteen million public school students raised over $3 million in aid, sewed and knit garments, rolled bandages, and sent "friendship boxes" containing gifts and school supplies to war victims and orphans overseas.[52] In response to their efforts, the young European recipients sent token gifts, such as handkerchiefs and dolls, and letters of gratitude that told the Americans about how they benefited from the gifts.[53] The resulting flurry of friendly correspondences between children led the JRC to formally organize the International School Correspondence Program (ISCP) as one of its peacetime efforts in 1922.[54] While Americans would continue to send relief supplies to countries in need, students in all countries with JRC chapters were now encouraged to exchange handmade "albums," also called scrapbooks or portfolios, containing letters, artwork, photographs, and information about their countries, as a form of cultural and educational exchange.

As the JRC expanded into new countries after the war, growing to a worldwide membership of eighteen million by 1937, the ISCP expanded along with it. The network of participating schools grew from forty-six countries in 1925 to fifty-five by 1929. In 1926, over five thousand school groups in the United States were participating in the program, involving at least a quarter million students.[55] American children exchanged albums with peers in dozens of countries, including England, France, Germany, Italy, Switzerland, Romania, Lithuania, Bulgaria, Turkey, Haiti, Mexico, Estonia, New Zealand, Colombia, Paraguay, Cuba, Spain, Belgium, Poland, South Africa, Syria, Japan, and others. A map of the ISCP's network prepared for the League of Nations Secretariat celebrated the global reach of the program, highlighting the 259 different combinations of exchanges that had taken place between participating countries. The United States, recognized as the originator of the ISCP, was clearly the dominant participant in the program, with ties to virtually every other country. But in a noted improvement within the international correspondence movement that had taken root in Europe and the United States at the turn of the century, this program enabled countries that were traditionally excluded from Euro-American exchanges, such as China, Poland, Brazil, and Italy, to correspond directly with the Americans, Europeans, and each other.[56]

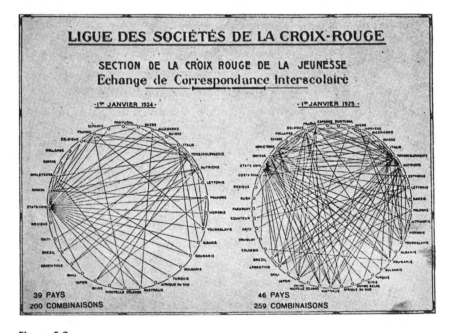

Figure 5.2
Map illustrating the international exchange of albums coordinated by the Junior Red Cross. In Spencer Stoker, *The Schools and International Understanding* (Chapel Hill: University of North Carolina Press, 1933), 215.

Educators' discourses about the program revealed a widely shared optimism about its potential to amend international hostilities and promote a new outlook of "world citizenship" among youths.[57] For Arthur William Dunn, the leader of the community civics movement who went on to serve as national director for the American JRC after the war, the ISCP was unrivaled in its ability to teach democratic ideals of cooperation and service around the world. "Each pupil and teacher engaged is linked up with ten million other teachers and pupils, not in correspondence alone, but in common social enterprise. So that the civic values of post-war work have been perpetuated and firmly established."[58] As the prominent English writer and war correspondent Philip Gibbs testified in the *Journal of Education* in 1921, the JRC's relief work and budding ISCP effort represented "the most beautiful experiment in the reshaping of human society which has happened since the world went mad in 1914." Believing that only children possessed the innocence and openness of mind to overcome the "innumerable frontiers of prejudice and ignorance" perpetuated by older generations, Gibbs proposed that building "a world-wide League of Children on this model" might be more effective at promoting peace than the fledgling League of Nations:

> All our discoveries in science, our mastery over mechanism, which seemed to promise a new and better world, have only made slaughter more widespread, the killing business more efficient, the communications and challenges of hate more rapid. . . . Is there any remedy? . . . I confess again that I do not see much hope from the men and women who are present responsible for the state of the world. . . . The hope of the world is in the younger generation. . . . This scheme of the Junior Red Cross of America is inspired by that vision of moral training which may lead us out of the dark jungle of old passions between nation and nation, and people and people. I like its simplicity and its practical plan. I like its appeal to the imagination of childhood and its encouragement of things made by their busy hands.[59]

For many observers, the scrapbooks created and exchanged through the ISCP were visual testaments to the liberal postwar ideals of international cooperation, communication, and goodwill. Whereas the Peabody program emphasized *individual correspondence*, the ISCP specialized in *collective* or *group correspondence*, or the exchange of single scrapbooks produced collaboratively by an entire class or school.[60] Each scrapbook was unique, and included a heterogeneous mix of items such as student-authored letters, essays, poems, photographs, newspaper and magazine clippings, drawings, pressed plants and natural specimens, coins, textiles, handiwork, and other materials affixed to paper with tape, string, or glue.[61] As projects that could take weeks or months to complete, the scrapbooks were often meticulously and elaborately constructed with the aim of giving their receivers a panoramic view of home, school, community, and national life.[62] As Dunn observed, this "co-operative effort" of group correspondence allowed for the "maximum extension of influence" of its lessons in democratic citizenship, as students would learn how to work together and integrate multiple academic subjects in their effort to produce a collective portrait of their community.[63]

Some schools exchanged a single pair of scrapbooks while others engaged in extended correspondences that lasted several months or years. As an unusually direct and visual medium of peer-to-peer international communication, participants often posed questions to their fellow students abroad, eagerly awaited their responses, and pored over the scrapbooks and letters they received in return. One class in Arizona, on receiving a mysterious

Figure 5.3
Junior Red Cross albums from Hawaii, "brilliant with colored pictures and weaving." Ellen McBryde Brown, "Interschool Correspondence Promotes International Understanding," *School Life* 13, no. 9 (1928): 173.

piece of coral along with an album from the Virgin Islands, immediately sent nine questions about it, including whether it was poisonous, noting they were "very anxious to know more about it." Some students wanted their partners overseas to confirm or elaborate on information about their country that they had gleaned from popular media, such as films or books. As a class in France wrote to one in Elgin, Illinois, "What sort of films do you like? Will you please tell us about the life of the Indians? Will you explain your most popular sport? Can you send us pictures of your National Parks? We should like to know about the greatest men in your history."[64] Other children offered poems and artwork that testified to the sheer fascination and thrill of being in contact with far-flung peers—information that the JRC proudly published in its promotional literature. When students in Brazil wrote to a class in Minnesota, they poetically chronicled the joy of communicating across the hemisphere, and developing a sense of connection to a distant people and places:

> It was with pleasure that we received a dainty album made by you, as it increases still more the cordiality between the children of the two countries. . . . Minnesota is above the equator, and therefore there is a contrast in our lives. The Sun, being only one, and owing to the roundness of the earth, cannot be agreeable to us at the same time. When he visits you most intimately he sends us only pale rays, or when we are enjoying the beautiful days of spring you there are in the sad days of autumn, and when we enjoy the delights of summer you have the dark days of winter. Our vacation is in December and yours would be in June. But if nature thus places us in opposite positions, the Junior Red Cross extends its banner over the Americas to unite our hearts.[65]

In promotions for the correspondence program, American JRC officials touted it as an exercise in mutual enlightenment, though the kind of enlightenment obtained by the children depended on where participants lived and how developed their societies were. Similar to discussions of educational motion pictures, advocates commented on how scrapbooks could contribute to the "civilization" of peripheral nations while revealing to American children how civilized the rest of the world already was or was becoming. Teachers reported that on seeing snapshots of peers in other countries, American students were surprised to encounter "splendid looking, intelligent boys and girls of other lands" who shared, according to one Kansas educator, "the same hopes, aspirations, love of home, ideals and loyalties as they themselves."[66] Promoting the program to American teachers in *School Life*, the official publication of the US Office of Education, in 1928, one JRC official noted how a high school for boys in the British colony of Kenya wrote to a school in Portland, Oregon, using "excellent English, not one word of which the writers knew six years ago." Contrasting the modernity of the ISCP and colonial education system against the traditional culture of the young letter writers, who were members of the Kikuyu tribe, the JRC official marveled that some of these "high-school students whose fathers go about in skins and carry spears" could now share with the world their accounts of hunting, building huts, gardening, and making weapons and ornaments. "It seems safe to say," the official wrote, "that a whole stack of books about the Kikuyu could not begin to give the Oregon children such a vivid impression of them as these letters and sketches coming from right under the Equator and from the pens of Joseph Mwangi Simeon, Hodea Wauiaina, Hagal Duolo, Elind Wambu, and the rest."[67] Accounts like these suggest that while the ISCP was valued for broadening the cultural horizons of Americans, it was equally valued for contributing to non-Westerners' march toward modernity, and demonstrating how the Western educational and communication systems were contributing to global social progress.

As more schools signed on to the program in the late 1920s, the ISCP quickly became a centerpiece of interwar efforts to teach world-mindedness in American schools. Studies of the program confirmed its effectiveness in molding young people's social attitudes. In 1935, Charlene Rector, a teacher of social studies in a middle school in Muncie, Indiana, tested how an ISCP scrapbook from Japan shaped her students' attitudes about the country and then shared the results with Indiana University researchers. Before presenting the scrapbook to the class, Rector first surveyed her students' attitudes about the Japanese, and discovered that they held a number of negative stereotypes based on movies, stories, and "the few Japanese they had seen." Students believed that the Japanese "like to fight," "like flowers," "are sneaking," live in "tumbled-down shacks," "are nervous," and "think their emperor is God." After surveying her pupils, Rector then instructed them to study the Japanese scrapbook, which contained snapshots of its student authors, descriptions of national holidays and recreational activities, artwork, and other objects including "mounted seaweeds, moths, and butterflies" that were "minutely and accurately painted in water colors." Afterward, Rector surveyed her students again and found that their impressions of the Japanese had changed dramatically. Students now described the Japanese as "patient," "painstaking," "intelligent," "calm," "self-controlled," "patriotic," and "artistic." Through this miniature study in media effects, Rector concluded that JRC scrapbooks could transform children's negative attitudes about "Orientals" and other cultures. As she summarized it, the scrapbook, with its diverse, student-made texts and materials, had given students "an entirely new and understanding viewpoint of these formerly 'queer' people."[68]

Further evidence of the socially transformative power of the album exchange appeared in a doctoral thesis by Everett Baxter Sackett at Columbia University in 1931. In an effort to measure the program's efficacy in teaching international understanding, Sackett studied eighty-two albums exchanged between students in eighteen countries, along with questionnaires filled out by teachers who had participated in the program during its first decade. Ranking the album contents and teacher comments according to their apparent cultivation of "good will" or "ill will" between nations, Sackett concluded that the exchanges of scrapbooks were mostly effective in fostering international friendship, even if, in rare cases, they also caused "friction."

In terms of promoting goodwill, Sackett noted that albums were almost always stocked with positive gestures of friendship, such as statements of welcome and questions about life in the other country. They were typically received by schools with tremendous enthusiasm, publicly displayed in exhibits and libraries, and "thumbed through by many hands." Like pen pal correspondences, the album exchanges promoted the sought-after progressive ideal of learning through activity, as they inspired students to scour newspapers for information about their correspondents' countries, create lantern slides out of the albums' contents, plant gardens out of seed packets tucked inside their pages, and attempt to imitate the drawing and handiwork techniques of their foreign peers. In contrast to pen pal exchanges, Sackett asserted that the group correspondence of the ISCP promoted communications that were more scholastic than personal in nature, and thus "minimized the individual." Still, in contrast to a textbook, they conveyed a more "human" picture of a country. Participants experienced "a realization that people of other lands are really 'human,'" he wrote. Even when the occasional album contained "no more personal interest than the usual textbook, . . . the fact that the material is from a foreign country and is unusual" was enough to arouse students' interest in ways that textbooks or other mass-produced materials would not be able to accomplish.[69] In other words, by virtue of their handmade construction and foreign origins, the JRC scrapbooks were uniquely effective devices for revealing the "human" qualities of faraway peoples and thus promoting international understanding.

The instances in which scrapbooks caused "friction," on the other hand, typically resulted from moments of miscommunication, disruptions in shipment, or larger geopolitical tensions between the corresponding countries. Schools reported "ill feeling" when partner schools were slow to respond to their albums, neglected to answer questions posed in earlier correspondences, or did not make an effort to match the quality of albums that preceded them. Such feelings could translate to negative attitudes toward the nation at large. Additionally, Sackett noted that some albums contained unwanted signs of nationalist and political ideology, such as contents that glorified war leaders, pictures of battles and armaments, and references to international trade and territory disputes. The Red Cross Societies Secretariat, which claimed to screen all the albums before forwarding them to schools, made efforts to remove or censor such material, but accounts suggest these efforts were inconsistent.[70] One glaring instance occurred in 1942, shortly after the United States entered the Second World War, when students at Tucson High School in Arizona received a scrapbook from a school in German-occupied Austria that was replete with swastikas. The American school officials promptly denounced the scrapbook as a missive of enemy propaganda. "Every photograph in the entire album testified so obviously to the imposed presence of the swastika," wrote the school dean, "that the portfolio was quickly laid aside by pupils and teachers alike, with a feeling of utter disgust."[71] The inclusion of such material in some JRC scrapbooks highlights how easily geopolitical ideologies and expressions of nationalism could be reproduced in the intercultural communications of children, replacing feelings of amity and goodwill with suspicion and hostility.

In addition to being hampered by occasional displays of jingoism, the ISCP's aim to foster equitable cultural exchanges among nations was often in tension with the humanitarian mission of its parent organization, the JRC. The charitable spirit of the program was particularly pronounced in the United States, where JRC members pledged to promote "worldwide friendship" through "service to others," with an emphasis on reducing human suffering abroad and within their own communities. Throughout the 1920s and 1940s, promotional materials for the ISCP reminded participants of its origins in the noble relief efforts led by American children during the First World War. In 1928, Dunn described the album exchange that resulted from those efforts as a way for American children to "share, in a very fundamental way, in post-war reconstruction" by sending "work of their own hands and sacrifice of their own accustomed pleasures." Scrapbook production, in short, was ambiguously positioned as an act of both two-way educational exchange and one-way volunteer labor in which fortunate American children provided aid as well as enrichment to their less fortunate peers abroad. The language of personal sacrifice and altruism continued to guide the program during World War II, when the American JRC experienced an influx of new members. Members of JRC chapters across the country continued to exchange scrapbooks with children overseas while also raising money for war victims and preparing, as in the First World War, "gift boxes . . . for less privileged children in foreign lands." Calanthe Brazleton, the dean of girls at Tucson High School, praised her students' "unselfish attitude" in doing this work in wartime, and noted that because the fruits of their labor would be likely to be received by children who were "in no condition to reply or send gifts in exchange," the local Red Cross office had organized a special turkey dinner to honor their efforts.[72] Through these sorts of activities and discourses, the JRC blurred the line between the equitable exchange of educational materials and unilateral shipment of material aid, reinforcing problematic associations between cultural exchange, economic development, and ideologies of American paternalism that would become more pronounced during the postwar period.[73]

These tensions were evident in a JRC scrapbook, a series of pen pal letters, and other international texts created by American students for the Children's Crusade for Children (CCC), a coordinated fund-raising campaign held in schools across the United States during the week of April 22, 1940, to aid children displaced by the war in Europe. Organized by the educational reformer and social activist Dorothy Canfield Fisher with the support of Studebaker, the US commissioner of education, the CCC encouraged students to donate "as many pennies as they are years old" in collection cans distributed to their classrooms. With the help of volunteers from the JRC, Boy and Girl Scouts, 4-H Clubs, and other youth organizations, the cans were then turned over to local banks, which forwarded the funds to the national campaign for redistribution in Europe. A massive publicity

Figure 5.4
Promotional images for the American Junior Red Cross linked humanitarian aid and assistance to international understanding. Douglas Griesemer, "Shop Talk: International Correspondence of the Junior Red Cross," *Elementary English Review* 16, no. 7 (1939): 284.

effort, including radio, newspaper, and comic book advertisements featuring the endorsements of Eleanor Roosevelt and Pearl S. Buck, rallied children and educators to the cause. In addition to raising money, schools were urged to "dramatize the multi-sided aspects of this educational campaign" through lessons and activities about the importance of service and sacrifice in promoting democracy as well as international understanding. Across the United States, schools used the weeklong campaign as an occasion to engage in a number of international activities, including putting on pageants and assemblies, producing exhibits and bulletin boards of foreign costumes and dolls, and writing to pen pals. The hope was that these activities would not only aid in the fund-raising effort but also "give our American children an opportunity to make, themselves, a generous gesture of friendliness, direct to other children" and thus "help counteract the negative impression of overwhelming evil in human life."[74]

A recurring theme in CCC materials, including those officially disseminated by the campaign as well as those unofficially produced by students and teachers, was that American children had a duty to help their peers in war-torn Europe, and "rescue" them with their donations and goodwill. The rescue theme was vividly portrayed in the campaign's official logo, created by the artist Normal Rockwell, and reproduced on posters and collection cans. In it, a smiling, golden-haired American boy stood tall and at the ready, with one hand digging into his pocket, eager to make a contribution to the campaign. Behind him, huddled in the background, were three haggard children—war victims—with sacks slung over their backs, looking wearily into the distance.

For their part, children reproduced the Rockwell scene and its message of American altruism in countless drawings, pen pal letters, skits, and assemblies as well as on bulletin boards. One photograph of a children's tableau is suggestive of a trend that was common in school pageantry at the time, in which children with fair hair and light skin were given the role of the "America" figure while their darker-complected classmates were assigned the roles of downtrodden refugees. In a spelling bee held in honor of the CCC, children in one school were challenged to spell the words "refugee," "starvation," "suffering," "contribute," and "volunteer." In another, children wrote to their pen pals about the purpose of the campaign. "Miss Rose told us that we ought to sacrifice something of our own instead of asking our mother or father for money, because we have so many pleasures and interesting things to do," wrote Geraldine Woods of Penn Yan, New York, to her pen pal, Ruth, in an unspecified country. She surmised that the children of other countries "probably wouldn't have enough food or homes of their own if it wasn't for the people of America who feel sorry for them and want to help them."[75]

Figure 5.5
The official poster of the Children's Crusade for Children (1940), created by Norman Rockwell and modified by a school in Oshkosh, Wisconsin. Children's Crusade for Children Records, 1939–1940, Library of Congress, Manuscript Division.

In a remarkable scrapbook created in commemoration of the CCC by members of the JRC chapter in Hoboken, New Jersey, students directly addressed the "children of Europe" who would receive their aid with poems and drawings detailing their own good fortune and helpfulness. "Here

Helping Hand from Uncle Sam's Children

Figure 5.6
Children of the Rochester Public Schools in Rochester, New York, interpret the Children's Crusade for Children. *Rochester Times-Union*, April 26, 1940, in Rochester Public Schools Scrapbook, box 23, folder: Publicity, Library of Congress, Children's Crusade for Children Records, 1939–1940, Manuscript Division.

come our pennies, nickels and dimes," wrote one student in a colorful drawing of US coins traveling across the Atlantic Ocean. On a page titled "GREETINGS," a child named Doris penned an original poem expressing Americans' sympathy for their less privileged peers abroad:

Dear little children in Europe,
I am very sorry to hear
There is so much trouble in your country
And you have so much to fear.
I wish you were here in America
Where everyone is fine and dear.

> We in America will help you
> You have nothing more to fear.

On another page, alongside an envelope addressed to "kiddies of foreign lands," another student explained to the foreign reader that acts of giving and charity were an essential part of being American:

> There's nothing like a letter
> To bring others good cheer—
> 'Specially when it comes from tots
> Who're contented over here—
> Bringing money that buys food
> For those kiddies far-away—
> It's an "old American custom,"
> So just start right in to-day.[76]

In a similar poem, "The American Way," a child included an image of happy, healthy American children at play—providing a contrast with other images in the album of war refugees seeking food aid and medical care. Her poem read:

> Here are two American kiddies
> A playin' in the sand
> With ne'er a worry about blackouts
> 'Cause they live in a peaceful land.
> Pity the poor little foreigner
> Whose shovels mean work in the fields.
> Let's call it a special duty
> To supply them with just a few meals.

Altogether, the texts created for the JRC and CCC were at once ambitiously internationalist, urging a greater sense of responsibility among American youths to serve the community of nations, and deeply patriotic and paternalistic, celebrating the United States' privileges of wealth, stability, and power. As a CCC promotional pamphlet reminded educators, in addition to providing aid to war victims, the effort would "aim to inculcate a vivid consciousness in American children of the unrecognized blessings they enjoy in this land" and would thus serve as a "continuing lesson in living patriotism." Children dutifully echoed these themes in their own artwork and writings about the fundraiser. "As we give we should think of the thousands in Europe that are dying and starving," remarked one eighth grader in his school newspaper in Rochester, New York. "We have never suffered any of this and we rejoice that people in this country live in perfect harmony."[77] On the other side of the country, a high school newspaper in Helena, Montana, urged students to give to the campaign "if you are grateful for having been born into a nation at peace, if you are thankful that all the years of your lives have been spent in a democracy that waged no wars and repelled no invasions—a government which has permitted the greatest degree of individual liberty ever known to man."[78] While a few young people voiced doubts about the CCC—particularly the idea of sending aid abroad when poverty and other problems existed at home—it was much more common for children to echo the official refrain that even the poorest American children were "rich by comparison" to their peers in Europe.[79]

By participating in service activities like the ISCP and CCC, American children learned that engaging in cultural exchange with the international world could also involve acts of charity and personal sacrifice. A recurring theme in these programs was that US youths should feel grateful to live in a secure, democratic country while their peers abroad suffered under oppressive or authoritarian regimes. Like other lessons on world citizenship from the era, these activities presented a largely self-congratulatory and Euro-Amerocentric vision of cosmopolitan citizenship that ignored the existence of widespread racial inequality within the United States. Through these lessons, white American youngsters could expand their knowledge of the international world while contemplating other nations' struggles toward achieving a level of democracy, wealth, and stability that they already enjoyed in the United States.

Tokens of Friendship and Power: The CWFC's Object Exchanges

While thousands of American children authored letters and scrapbooks to exchange with peers abroad, others sent token objects as a material symbol of their nation's goodwill. Unlike the international correspondences arranged by the Peabody Bureau and JRC, these exchanges relied not on letters or student-made texts but rather on onetime shipments of material items such as dolls, school supplies, and toy treasure chests to children in a single country. While conducted in the name of international friendship, these shipments were also regarded by their organizers as a form of material aid, and likewise conveyed information about nations' varying levels of economic wealth and infrastructure. Initiated by American Protestant missionaries and peace educators, they served as much as tokens of friendship and goodwill as reminders of the geopolitical relations and economic inequalities that stratified nations and colonial territories in the post–World War I era.

The program that most clearly encapsulated these ambiguities was the CWFC's Friendship Project, founded in New York City in 1926 by Sidney Gulick, an American missionary to Japan as well as vocal advocate of improving Japanese-American relations. Dismayed by the rise of anti-Japanese rhetoric and passage of the Immigration Act of 1924, Gulick believed that international understanding needed to begin with reforming the attitudes of young people, remarking, "We who desire peace must write it in the hearts of children." From 1927 to 1950, his committee orchestrated a series of national campaigns to get American schoolchildren to prepare and send symbolic objects of goodwill to peers in different countries and US-occupied territories. Despite the program's origins in missionary work, the CWFC coordinated its efforts with both religious and secular organizations, including public and private schools, the Camp Fire Girls, YMCAs, the Boy Scouts, and the JRC.[80]

While its 1927 shipment of thirteen thousand "friendship dolls" to Japan was the first and most widely publicized of its projects, the CWFC went on to conduct several more campaigns on an even larger scale, including sending thirty thousand schoolbags to Mexico in 1928, twenty-eight thousand treasure chests to the US-occupied Philippines in 1929–1930, six thousand treasure chests to US-occupied Puerto Rico in 1931, and twenty thousand albums, or "friendship folios," to China in 1932.[81] The CWFC members selected the destinations and token objects (also referred to as "symbols") based on their assessment of which peoples were most in need of "an assurance of our friendly interest and our goodwill," and which objects would carry "special meaning" for the children of that place.[82] Children were encouraged to raise small amounts of money to order the gifts—either directly through the CWFC or according to their approved guidelines—and then personalize the items with their own "goodwill messages" and accessories before sending them back to the organization's headquarters for shipment abroad.

The CWFC projects offer insight into how interwar missionaries and peace educators conceived of toys and material objects as potent instruments for building international understanding among children. When planning the doll exchange in 1926, for example, the committee agreed in meetings that "the dolls are not to be sent as gifts, with the idea of giving something to Japan, but they are to be representatives of the children of America, bearing messages of friendship and goodwill."[83] To play up the notion of the dolls as living messengers rather than dormant objects or playthings, the CWFC launched a public relations campaign—"Say It with Dolls"—to portray the dolls as pint-size emissaries of the United States. After securing a miniature passport, visa, and travel ticket for their dolls from the CWFC headquarters (temporarily named the "Doll Travel Bureau"), children were instructed to creatively outfit their dolls with regional clothing and luggage, attach a written "message of goodwill" and return address signed by each child, and involve their communities in hosting "a 'farewell party' to say 'goodbye' and to wish the doll 'bon voyage' as it starts its long journey and success in delivering the message."[84]

Though subsequent CWFC projects featured nonanthropomorphic objects, such as schoolbags and treasure chests, the program officials described them all in various turns as "ambassadors" or avatars for the children who sent them as well as vessels of their heartfelt goodwill. In preparation for the Friendship Project to Mexico in 1928, for instance, the CWFC chose schoolbags as the symbol of friendship "to carry the good wishes of the children of America to the children of Mexico." The handsome bags, manufactured in the United States with Fabrikoid, a synthetic leather used in automobile upholstery, featured an embossed message of "Goodwill Greetings" in Spanish and English on the exterior, framing a playful image of children holding hands. Each bag contained a set of coloring cards with pairs of iconic symbols referencing the revolutionary histories of the United States and Mexico, including George Washington and Benito Juarez along with the American and Mexican liberty bells. Children were encouraged to personalize the bags with their own token gifts, including toys such as jacks and puzzles, and hygienic supplies such as soap and toothbrushes.[85]

The notion that material goods could ferry sentiments of goodwill across oceans and borders was the organizing principle behind each CWFC project, and according to its reports, was echoed back by the foreign officials and children who received them. At a parade in Manila celebrating the arrival of twenty-eight thousand friendship treasure chests filled with books, toys, and school supplies sent by the Americans, the city's mayor, Tomás Earnshaw, stood beside a float with a giant model of a chest and delivered the following remarks to thousands of Filipino children:

> More than simple gifts, the toys in these chests should mean to you a message of love, friendship and goodwill. Gathered by loving hands from cities throughout the U.S., and packed with solicitous care for the long voyage across the wide Pacific that they may safely be delivered to you, each one of the seven hundred fifty thousand little toys contained in these chests is a cordial handclasp, an affectionate embrace, as it were from American children to Filipino children.[86]

Elaborate reception ceremonies for the CWFC tokens were presided over by prominent officials in each country, including the imperial family in Japan, the president of Mexico, and ambassadors and educational authorities. The participation of high-level officials in these events lent further legitimacy to the notion that token exchanges among children were effectively an extension of state diplomacy. It was an idea that would later undergird the "sister city" or "city-twinning" initiatives that were a centerpiece of the People-to-People Program during the Cold War, when municipalities in different countries formed and commemorated symbolic partnerships through the exchange of visitors, monuments, and public art.[87]

Accounts detailing the reception of the gifts abroad suggest that the traveling objects served their purpose in extending, albeit perhaps momentarily, feelings of goodwill. Each receiving nation hosted events for children in multiple cities to view the gifts, with some receptions reportedly involving tens of thousands of children. In schools and ceremonies, observers described the children as "beside themselves with eagerness and joy," and thrilled beyond description to receive the materials. Many children used the return addresses attached to the objects to send letters of thanks the United States, initiating countless correspondences and pen friendships. Observers noted that for the young receivers, the significance of the materials lay not only in their novelty or intrinsic value as playthings but also in the idea that they had traveled great distances and with friendly intentions from the children of America.[88] In a message of gratitude to the CWFC reported by the *New York Herald Tribune*, Moisés Saenz, assistant minister of education in Mexico, described the lengthy journey made by a single CWFC schoolbag to reach the remote Tiburón Island, where indigenous children reacted with such enthusiasm that they immediately worked to reciprocate the gesture using the meager materials they had on hand:

> The bag was carried eighty miles across desert by auto, then by car, horseback and rowboat to Tiburón Island, "then on the back of an Indian for several miles to the Indian camp, where the bag was opened and the 'good will' things distributed

to the Seri children, who were made very happy to receive them, as they rarely receive any gifts, except persecution and ill will. These primitive people, wild as coyotes, appreciate kindness, generosity and thoughtfulness. A number of children immediately went along the shore of the gulf and one hundred feet distant and in half an hour returned with lovely shells washed up by the surf. These are sent [to] you as a small token of appreciation from one of the wildest tribes in America."[89]

Such accounts testified to the symbolic power of traveling objects—whether elaborate schoolbags or humble seashells—to build a palpable sense of connection among far-flung children. At the same time, Saenz's comments about the Seri children being "primitive" and "wild" people in need of the "kindness" emblematized by the American schoolbag highlight the salience of humanitarian and civilizing discourses in the CWFC projects. By mobilizing young Americans to send toys and supplies to children overseas, the CWFC worked in alignment with the JRC, missionary networks, and other service organizations to promote sanitation, education, and economic growth in developing nations. As Julia Irwin has argued, the economic as well as material aid dispensed by such organizations became a key mechanism through which Americans interacted with the wider world and understood their country's role in international relations in the twentieth century.[90] In a more troubling interpretation, the inclusion of hygienic supplies as a symbol of goodwill evoked a colonial tradition of British and American depictions of soap as a civilizing, "whitening" agent that could scrub away colonized people's blackness.[91] Despite the committee's professed attempts to strip its projects of any "missionary flavor," and package the outgoing materials as messages of equitable friendship rather than gifts or aid, their problematic contents and mass, unsolicited shipment overseas conveyed a subtext about the economic might of the United States, and presumed neediness of recipient nations and territories for "assurance" of its generosity along with modern products and supplies.[92]

Indeed, the CWFC encouraged American children to prepare their tokens in a manner that aligned with the assumption that the United States had a responsibility to help or at least to model modernity for the people of other industrializing nations. Children sending dolls to Japan, for example, were instructed that their dolls "should be new and should be carefully dressed in every detail, since they will serve as models in a country where habits and customs are undergoing rapid changes." Tellingly, in order to model modernity, the CWFC decided that it was important for the dolls to be white. In its internal discussions about whether to allow children to send "colored" dolls, the committee decided to include instructions in the official literature that stated that the dolls should "look like attractive and typical American girls," which would "indirectly suggest that the dolls should be white."[93] Such assessments further underscore how white reformers' efforts to promote world friendship were commonly interlaced with the ideology of white supremacy, excluding African American children from both participation and representation in their programs.

Subsequent CWFC campaigns encouraged children to pair their token gifts with basic hygienic and school supplies, such as drinking cups and soap in the Mexican schoolbags, books in the treasure chests to the Philippines, and vouchers to sponsor lunches from a local charity in Puerto Rico to accompany the chests that were sent there. From one campaign to the next, by coupling their messages of goodwill with specific materials intended to promote hygiene, literacy, and nutrition in the receiving countries, the CWFC Friendship Projects seemed intent on conflating a peer relationship of friendship with a paternal relationship of providing aid and developmental assistance.

Further reinforcing this notion of foreign subordination to the United States, the CWFC literature depicted the receiving countries as grateful to the point of being overwhelmed by the influx of generosity from America. A 1928 pamphlet publicizing the success of the Mexico project, for example, offered a stark visual contrast between images of American children preparing the schoolbags and Mexican educational officials receiving them. In the top row, photographs of white American children showcased the orderly assembly and ceremonial send-off of schoolbags from the United States. In the bottom row, in contrast, Mexican educational officials were depicted in a chaotic scene—captioned, "Receiving, Classifying and Allocating Friendship School Bags in Mexico City"—in which they appeared to be swamped by mountains of schoolbags delivered from America. The disorderly abundance of the schoolbags, heaped floor to ceiling in a facility in Mexico City, simultaneously suggested the tremendous and organized generosity of the United States, and lack of infrastructure in Mexico to adequately handle and distribute such aid. While the pamphlet urged the American readers, "Let us convince all Mexico of the real goodwill of the people of the United States," the project seemed to be less intent on commemorating friendship with the people of Mexico than it was about reasserting the image of the United States as a land of order and material abundance, and the dominant figure in US-Mexican relations.[94]

Despite the CWFC's aim to distribute tangible symbols of American goodwill around the globe, the material nature of its exchanges indirectly foregrounded the socioeconomic, technological, and infrastructural inequalities that existed between the United States, its occupied territories, and other countries. Some of the recipient nations and territories responded to the American gifts by sending their own "goodwill greetings" in return. Japan sent fifty-eight ornately handcrafted dolls (one from every prefecture and major city), Mexico sent forty-nine exhibits on traditional arts and handicrafts (one for every US state and the District of Columbia), and the Philippines organized a group of schoolchildren to transmit a message of thanks to the United States over the radio. While the CWFC eagerly arranged for these gifts to be exhibited around the United States in schools,

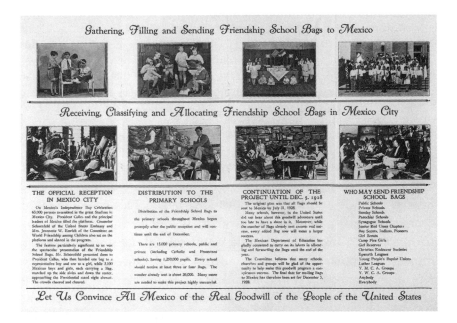

Figure 5.7
Committee on World Friendship among Children, "Friendship School Bags to Mexico," 1928. Federal Council of the Churches of Christ in America Records, 1894–1952, box 44, folder 21, Presbyterian Historical Society.

libraries, museums, and department stores, the notable difference in the scale and organization of the projects was nevertheless apparent in the relatively small number of objects that arrived from overseas.[95] As an exchange of gifts, the demonstration of material generosity was noticeably on the side of the initiating country, the United States.

Still, the gifts sent in return provided an opportunity for the recipient countries to speak back to their US counterparts and gently correct what they perceived to be American misconceptions about them. In *Welcome to the Doll Messengers*, a 1927 booklet published by the Committee on International Friendship among Children in Japan, a group formed in response to the CWFC doll project, the Japanese thanked the American children for the shipment of friendship dolls. Writing in the voice of the Japanese dolls it would send in return, it utilized the same concept of the doll as messenger to convey a rejoinder that subtly asserted Japan's modernity and status as a comparable industrial power:

> Do you know what kind of country Japan is? We are quite sure that you will say, "yes," but we . . . wonder if you really know the modern Japan. . . . We wonder what your doll-messengers said when they arrived in Yokohama after their long voyage across the Pacific. Perhaps they talked together in amazement and said: "Why, is Japan such an enlightened country as this? Before we left America we heard many things about Japan, but we did not believe that Japan was so far advanced in civilization. We were talking of the ancient Japan of Perry and his followers, and imagined it to be like the island empire of fairy stories. But this new country is like America. We shall never be homesick here."[96]

The passage highlights how countries on the receiving end of the United States' goodwill used tokens to convey their own messages and represent themselves on their own terms, reciprocating a gesture of friendship, on the one hand, while demonstrating their own national and cultural wealth, identity, and power, on the other hand. It also testifies to the larger point that the CWFC's token objects could be interpreted not simply as vessels of goodwill but also as oblique reminders of the advanced socioeconomic state of the United States in comparison to other countries. Though the tokens were welcomed and enjoyed by educators and children alike—and seemed successful in generating meaningful moments of intercultural contact and correspondence between children in the United States and other countries—they also carried a subtext of American superiority and power that reinforced the notion of the United States not as a friend or equal but instead a magnanimous "global parent charged with bringing up a family of nations."[97]

Mediated intercultural exchanges in schools were, by many accounts, deeply meaningful and memorable lessons. They gave scores of American students a personalized perspective on foreign cultures through unprecedentedly direct communications with far-flung peers. Originating at the turn of the century and exploding in popularity after World War I, exchanges of pen pal letters, radio greetings, scrapbooks, and token objects reflected a rising American interest in media-enhanced instruction, participatory learning, and internationalism, and a belief that a more peaceful future was possible if world friendship were "written into the hearts of children." Like the other technologies of global citizenship discussed in previous chapters, mediated intercultural exchanges were believed to transform children into tolerant, world-minded, yet patriotic citizens through controlled participatory and multisensory engagements with difference. In a manner that seems prescient of the promotional talk that surrounded the early World Wide Web, pen pal letters and JRC scrapbooks were held up as symbols of the utopian potentials of a more interconnected and networked planet. As such, they served not only world-minded educators but also government officials and telecommunications industries interested in advancing American connections to the world.

But while these practices were widely celebrated for breaking down stereotypes and building mutual understanding, their message of youthful amity was frequently undercut by their entanglement with larger nationalist ideologies and policies. Paradoxically, the activity of crafting and sending messages abroad in an expression of international friendship often doubled as an opportunity for children to rehearse and reinforce narratives about their own national superiority and exceptionalism. Openly embraced as a form of junior diplomacy, such exchanges frequently mirrored the dominant geopolitical priorities of the United States, and blurred the lines between equitable practices of exchange and paternalistic policies of providing American aid, assistance, and advice to the rest of the world.

6   A "Classroom without Walls": On the "Global Village" and Promise of Educational Technology

Today the globe has shrunk in the wash with speeded-up information movement from all directions.
—Marshall McLuhan, *Report on Project in Understanding New Media*, 1960

Today, the physical frontier is gone but a complex social frontier remains.
—Edgar Dale, "What Is the Image of the Man of Tomorrow?," 1960

A remarkable thing about claims that associate mediated learning with the cultivation of global citizenship is how evergreen they are. Over the past century, each new wave of information technologies has been accompanied by familiar promises that installing them in schools will "bring a universe of knowledge" to students, connect them with the wider world, and thereby usher in a more enlightened and democratic future.[1]

In this book, I have tried to show that educators, social reformers, media producers, and political leaders began collectively forging these utopian discourses in the first decades of the twentieth century, yoking an array of communication technologies and techniques in the classroom to an emerging ideal of world-minded American citizenship. Interested in legitimizing new media for school use, and fulfilling progressive visions of the school as an active, multisensory, and civic-minded center of learning, together they cast a host of sensory aids and activities—including new screen and sound media, school museums and scrapbooks, pageants and student assemblies, and pen pal exchanges—as "devices" that could "bring the world" to children and prepare them for democratic citizenship in an interconnected age. Hopeful that the era's communications revolution could be enlisted into the service of not just popular amusement or commerce but education and peace too, reformers established ways of talking about and applying new sensory apparatuses in the classroom to develop in children a generalized appreciation for cultural differences that would not challenge the idealization of white, middle-class US citizenship. As the United States embraced an increasingly powerful role on the world stage, these uses of media revealed tensions between a well-meaning interest in cultivating "world friendship" at home and abroad, and a strategic interest in ensuring the continued dominance of white, Anglo-Saxon cultural values in the United States along with the growth of American expansionism, economic development, and influence around the globe.

This kind of strategic cosmopolitan thinking is still evident in descriptions of educational technology in the digital age. Since its commercialization in the 1990s, the internet has been cast as the latest and most sophisticated tool to help Americans virtually see, understand, and lead in the globalized world. "Many people dream about traveling around the world, but few get the chance to do it," observed *Internet for Kids*, a 1996 educational guide to the World Wide Web. "On the Internet, you can be a lucky traveler in a unique way. You won't be taking a plane or car or ship to faraway destinations. Instead, you'll use your computer and modem to meet people in foreign countries, ask questions, and experience other cultures without even leaving home." In the introduction for the book, Democratic senator Bob Kerry of Nebraska noted that connecting American kids to the internet was critical for their future economic prospects and informed citizenship. "Telecommunications technology remains vital to our children and our future. The computer and modem and the infrastructure that connects them with the rest of the world are as indispensable to our children's education as the textbook, and it is up to us adults to see that our children have these tools."[2] The book's publication coincided with the passage of the Telecommunications Act of 1996, an initiative of the Clinton administration to loosen regulations on the telecommunications and broadcasting industries to promote, among other things, increased competition on and access to the internet. When he signed the law, President William Clinton was keen to highlight that schools would be among its primary beneficiaries, asserting that the "new markets" it created would "help connect every classroom in America to the information superhighway by the end of the decade."[3]

In addition to helping young Americans *compete* in the global economy, the use of new technologies in school would presumably expand their cultural horizons and shape their social attitudes, making them more tolerant of cultural differences and thus better prepared to *cooperate* with people of different backgrounds and beliefs. President Barack Obama drew on this idea when in a highly anticipated speech in Egypt in 2009 aimed at mending US relations with the Muslim world, he pledged to "invest in online learning for teachers and children around the world, and create a new online network so a young person in Kansas can communicate instantly with a young person in Cairo." For American audiences back home, the ideal of cross-cultural cooperation was routinely couched in the language of national economic prosperity and geopolitical power. As US secretary of education Arne Duncan remarked in a speech to the Council on Foreign Relations in 2010, embracing opportunities for international education and exchange would make Americans "better collaborators and competitors in the global economy," and was "vital to our nation's capacity to compete, collaborate, and exert smart power."[4]

The recurrence of these claims in American political and educational discourse over the last century highlights how emerging communication technologies and practices have long been characterized as tools to not only instruct but also mold an idealized type of citizen that the nation and world needs. Equipping education with a diversity of media, the thinking has been, will likewise equip young Americans with the cultural and technological competencies needed to navigate a diverse future as well as advance national interests in a connected age.

How the Global Village Made American Global AV Mainstream

It is difficult to conclude a study of a discursive phenomenon as persistent and ever evolving as this one. One way is to briefly revisit the moment where it is commonly thought to have originated, in the Cold War discourse of a global village. In the 1960s, McLuhan, a Canadian communication scholar, became an international celebrity for his writings and media appearances in which he theorized the social impact of new media. Among the most famous concepts to come out of his best-selling books, *The Gutenberg Galaxy* (1962) and *Understanding Media* (1964), was that of the "global village," the notion that the proliferation of electric media in society (such as telephones, radio, film, television, satellites, and computers) was obliterating barriers to communication and extending humanity's powers of sensory perception, causing the world to "implode" into a more intimate and interdependent state.

> Today, after more than a century of electric technology, we have extended our central nervous system itself in a global embrace, abolishing space and time as far as our planet is concerned. . . . As electrically contracted, the globe is no more than a village. Electric speed in bringing all social and political functions together in a sudden implosion has heightened human awareness of responsibility to an intense degree.[5]

The concept quickly made headlines and had tremendous staying power, serving as shorthand for the accelerating globalization of culture, technology, and commerce that Americans saw as necessary to ensuring the triumph of free market capitalism over Communism. But nowhere did it gain more traction than in American education. By the late 1980s and 1990s, as the Soviet Union disintegrated, and the internet became available for educational and consumer use, the term appeared frequently in American educational literature as a descriptor for the increasing state of quotidian global connectivity and information exchange that was believed to be just over the horizon, if not already in place, in schools. For example, in a major 1998 study of educational technology called "The Neighborhood School in the Global Village," conducted with support from the National Science Foundation, Apple Computers, and the Japanese electronics producer Hitachi, researchers at Virginia Tech declared, "The classroom is no longer just a plain and well-worn cubicle inside a somewhat dilapidated brick building in the center of town; it is a node in a world-wide educational network."[6]

But as this book has suggested, the ideal of a globally connected and multimediated schoolhouse did not originate with McLuhan or the birth of the information society. As Charles Acland and Fred Turner have argued, McLuhan did not so much initiate Americans' interest in multimediated instruction as help "smooth the way for the full acceptance of technologized pedagogy" that had already been building in the fields of education, social science, and the arts for several years.[7] Moreover, this enduring interest in technology-enhanced learning, while perpetually cast in utopian terms about creating a more peaceful and democratic future, had long been tethered to self-interested anxieties about national power and defense. Notably, the first draft of *Understanding Media* was a report that McLuhan authored in 1960 as a consultant for the US Office of Education, in conjunction with the National Association of Educational Broadcasters. His report was one of the first to benefit from an influx of federal money for instructional technology research, allocated by the National Defense Education Act, signed by President Dwight Eisenhower in 1958 to strengthen education in science, technology, and foreign languages in response to the Soviet launch of Sputnik.[8] Written in McLuhan's characteristically postmodern style, the report warned that while the dawn of this "multi-media electric age" created an unprecedented opportunity for a more interconnected and interactive "global village," it also placed new demands on educators to move past the traditional reliance on pictographic and written approaches to instruction, and instead adopt a more "mosaic approach" to media in and outside the classroom. As McLuhan would theorize later in his career, the spread of new media was obliterating the divide between formal and informal education, transforming the world into a "classroom without walls." McLuhan's prose was unorthodox, but his message to defense-minded technocrats was clear: the US educational system could be either enhanced or rendered totally obsolete by the multidirectional flows of information outside the school grounds. To adapt to these new conditions, the teacher and student "must deal with all media at once in their daily interaction, or else pay the price of irrelevance and unreality," he warned.[9]

McLuhan's celebrity helped elevate into the national consciousness a global, multimedial, and technocentric vision of education, and yoke it to concerns about national security and competitiveness. But these associations had already been building in the educational community since long before World War II, and became firmly ingrained in its aftermath thanks to the postwar boom in the instructional technology movement. As scores of war-trained AV specialists returned to civilian life, they brought new energy, professionalization, and a defense-minded international outlook to what had been a scattered, experimental, and idealistic visual education movement. In 1945, the NEA converted its original visual education interest group, the Division of Visual Instruction, into a permanent department with full-time staff, separate conferences, and a new research journal, *Audio-Visual Communication Review*. In 1947, as mentioned earlier, it changed the name of the organization to DAVI, reflecting the increasingly multimodal orientation of the educational technology field. It quickly saw membership climb from a few hundred in 1945 to three thousand by 1955 and over nine thousand by 1970.[10] Like McLuhan, AV professionals believed that aggressively adapting the educational system to the new information landscape was essential to ensuring its success in the future.[11] Some drew from the growing field of mass communication research as well as the "integrated" view of organizations and technology associated with postwar "systems thinking."[12] While occasionally acknowledging Progressive Era principles of localized, sensory, participatory, and "active" learning, the postwar AV movement worked to make its instructional approaches more standardized, scalable, and exportable to national and global markets. Compared to the fledgling visual education initiatives of the prewar and interwar period, the postwar AV movement propelled educational technology to the forefront of the federal expansion of the public education system through closer collaboration with the booming technological sector along with an influx of financial support from the National Defense Education Act, NEA, and philanthropic organizations like the Ford Foundation.[13]

More than ever, postwar AV educators and advertisers cited the importance of technology in the classroom for preparing American youths for a more complex and interconnected world, now with the added subtext of supporting the nation's role as a global superpower, staving off the spread of Communism, and preventing nuclear war.[14] Aligning their efforts with space-age imperatives to boost American children's scientific, technological, and foreign language competencies, companies like Encyclopaedia Britannica promised to help teachers not just to "bring the world" but also to "bring the universe into the classroom."[15] Now, with the latest technology, American children would be free to virtually explore both the entire world and outer space.

While emphasizing technology's utility in advancing American power, AV boosters also noted its capacity to enhance international cooperation and promote global democracy. DAVI took an active role in internationalizing the instructional technology movement after the war, advising the United Nations, UNESCO, and World Confederation of Organizations of the Teaching Profession (WCOTP, also known as the "U.N. of the teaching profession") on establishing teacher training programs as well as international exchanges of AV aids and experts.[16] In 1961, with the WCOTP, DAVI published *Audio-Visual Aids for International Understanding*, a guide to over twelve hundred films, slide collection, filmstrips, and records from thirty-six countries. In the introduction, WCOTP secretary-general William Carr, a leading American advocate for world citizenship education since the 1920s, drew on the decades-old idea that educational media would aid teachers in "strengthening the bonds of comprehension and sympathy between the peoples of the world, and thereby advancing the cause of peace."[17] These efforts to globalize American AV dovetailed with other postwar initiatives by federal agencies—such as the Office of International Information and Cultural Affairs and the Division of Libraries and Institutes at the Department of State—to disseminate American films, books, teaching materials, and propaganda to foreign schools and libraries to promote a favorable image of the United States abroad.[18]

The work of the postwar AV movement brought into sharper relief the paternalistic and "civilizing" ethos of educational technology that had been present before the war, particularly in discussions about technology's potential to aid developing countries, low-income communities, and children of color. Even before the Soviet launch of Sputnik triggered an arms race of technology in education, domestic anxieties about "overcrowded" and "underperforming" American schools in the 1950s led to an intensifying interest in the mass-educational potentials of new technologies, especially television. Whereas pre–World War II visual education initiatives had been concentrated in white public schools, the outlawing of school segregation in 1954 heightened white Americans' awareness of the racial inequalities in the educational system, and prompted new efforts to harness technology to help Black and Latino youths "catch up" with their white peers. An interest in "compensatory education" helped to fuel the instructional television (also known as the educational television, or ETV) movement, which aimed to raise the quality of education, particularly in inner-city schools, rural areas, and developing countries, by broadcasting the lessons of a "television teacher" into classrooms. It also buoyed the programmed instruction movement, which called for using simple "teaching machines" and "programmed textbooks" to deliver sequenced steps of instructions, questions, and feedback to students on an individualized basis, and the emerging computer-assisted instruction movement. While promoted in the name of making education more equitable and efficient, many of these initiatives were criticized and scaled back in the 1970s for being too "mechanical," passive, dependent on corporate equipment and expertise, and removed from community and teacher control.[19]

These tensions were visible in one of the most ambitious experiments in American ETV, carried out in the unincorporated US territory of American Samoa in the mid-1960s. With funding from Congress and the backing of the National Association of Educational Broadcasters, the US-appointed governor of the islands, H. Rex Lee, sought to make television the primary mode of instruction in Samoan elementary and secondary schools. The Samoan ETV Project brought engineers, production crews, and dozens of television teachers—most from the mainland—to the islands to broadcast lessons out of new, state-of-the-art television studios. Samoan teachers, who were seldom involved in the planning or delivery of lessons, were retrained to prepare their students for the telecast lessons and lead follow-up activities after the broadcasts. At the project's peak in 1965–1966, most Samoan students spent between a quarter and a third of their class time watching telecast lessons.[20] The program at first drew international acclaim, attracting dignitaries and AV experts from around the world. President Lyndon Baines Johnson, on a visit in 1968, declared the program a success, stating, "Samoan children are learning twice as fast as they once did."[21] But the program quickly drew criticism from local teachers and parents concerned about an overreliance on TV and mainland technology. Samoan teachers, Larry Cuban later summarized, "wanted less beamed into the classroom and greater control over the lessons that were broadcast." In a fifteen-year study of the project headed up by the communication researcher Wilbur Schramm, researchers concluded that while ETV brought some educational benefits to the islands, it was not the "miracle drug" that many had hoped it would be. The project was scaled back in the 1970s.[22]

The Samoan ETV Project points to an imperialist impulse in American AV during the Cold War—an impulse that as the foregoing chapters have tried to show, was already ascendant in domestic imaginings of the role of American educational technology from the turn of the century. Under a generalized rhetoric of improving international understanding and equalizing opportunity for children at home and around the world, the visual and AV education movements assumed the inherent superiority—and universal desirability—of American-made technologies, techniques, and messages in achieving those objectives. The Samoan ETV Project also highlights an enduring disparity in thinking about how technology should serve different school communities according to race. In white schools, technologies were historically envisioned as supplementary aids to the curriculum that could help broaden students' cultural horizons as well as build up their civic reasoning and responsibilities. In minority and colonial schools, in contrast, technologies were either absent altogether, or envisioned as tools for basic, remedial instruction or, more troubling, "civilizing" nonwhite students. As Governor Lee reportedly remarked on the urgency of bringing instructional television to American Samoa, the community was "grossly undereducated" and "one that America needed to bring into the twentieth century in a hurry."[23]

This faith in the universal appeal and uplift of American technologies undercut the part of the AV movement's rhetoric that suggested that technology could foster better multidirectional communication and "mutual understanding" among cultures. For example, Edgar Dale and F. Dean McClusky, who remained leaders in AV education for decades after the war, often gestured to the idea that the global export of American educational technologies should be carried out in such a way that allowed for local contextualization, teacher control, and "two-way communication" between countries. At the same time, they openly assumed that educators in other countries would not only be eager to adopt American-style AV instruction but also would use it to make their societies more aligned with first world democratic and free market values. These ideas were evident in Dale's "Proposal for International Audiovisual Exchange," an address delivered to the annual DAVI convention in 1960. Speaking to an audience of American AV professionals, Dale proposed creating an international system of "clearinghouses" to facilitate greater circulation and exchange of AV materials, information, and personnel between countries. He said that while American AV educators should not enter such a project with the goal of "Americanizing" other nations, they should also not ignore the unique lessons and values—relating to education and government—that the United States could teach the rest of the world, particularly to the developing nations emerging from colonial rule:

> We do not want to impose our own government on anyone or indeed Americanize them. But our mood is one of great respect and pride for our governmental system. I do not suggest that we neatly package and export our political institutions yet I believe that they are worthy of continuing study for the light they can throw on the difficult governmental problems arising not only in the new nations of the world, but also in older ones, including our own. We Americans tend to forget our own remarkable history of self-government. . . . We would like to base our

exchange of ideas, techniques, and material goods on the premise that we want the people of the world to be different, to discriminate between alternatives, and to choose what they want. We do not want a homogenized world culture that is 99.99 percent pure. However, we do believe in the importance of a commitment to and involvement with those values deemed by critical thinking to be best.[24]

This view—that Americans should use technology to learn from other countries, but also already possessed the highest standard of technological and social development to which other nations would want to aspire— contributed to the growth of what Brian Goldfarb calls a *global educational mediascape* dominated by American technologies and ideology by the end of the century. Through campaigns like the Samoan ETV Project as well as other efforts to bring educational radio, television, and computers to developing nations, Americans established themselves as "benevolent leaders in introducing media technology—and hence knowledge and status in the global market—to underdeveloped regions."[25]

The United States' expansionist technological ethos continued to fuel ambitious campaigns to connect the world's classrooms through the end of the Cold War. The late 1980s and early 1990s saw the emergence of several American-initiated virtual exchange projects to support computer-mediated dialogue among schoolchildren as well as improve relations between countries in the West and behind the iron curtain.[26] Prominent among these was the New York State–Moscow Schools Telecommunications Project, launched in 1988 by the Copen Family Foundation with support from the US Department of State. The project connected students in a dozen classrooms in the United States and the Union of Soviet Socialist Republics via email and video speakerphones donated by Mitsubishi. In an electronic and digital revival of the tradition of mediated intercultural exchange described in chapter 5, students exchanged friendly messages over speakerphone and email, and collaborated remotely on shared school projects. Within a few years, other private foundations and schools launched similar programs to connect American and Soviet schools and universities—a trend that the *Los Angeles Times* cheered as a "Computer-Age *Glasnost.*" By the mid-1990s, the New York State–Moscow project evolved into what is now a leading organization for supporting virtual exchanges in education, the International Education and Resource Network (iEARN). With members in 140 countries, iEARN describes itself as the "largest nonprofit global network" for facilitating online educational exchanges. Shaping this grassroots mission,

Figure 6.1
"Audio-visual materials speak a universal language." F. Dean McClusky, *Audio-Visual Teaching Techniques* (Dubuque, IA: Wm. C. Brown Company, 1950), 107.

however, are iEARN's multiple partnerships with technology corporations, private foundations, and state and federal agencies.[27]

Since the turn of the twenty-first century, references to global and media-enhanced learning have become a taken-for-granted feature of educational and political discourse in the United States and other Western countries. Several notable efforts in this area have arisen through the grassroots work of nonprofit organizations, students, teachers, and researchers interested in exploring emerging communication networks as well as advancing intercultural learning and dialogue. Some of these continue to make use of pre-digital forms of intercultural communication and contact, such as the Flat

Stanley Project, in which children in different countries and geographic locations exchange handmade paper dolls in the mail, often while also engaging in computer-mediated forms of exchange.[28] Still, many of these efforts to teach global and technological competencies depend on funding and computer equipment from powerful philanthropic foundations, for-profit tech companies, and government agencies that have a primary interest in retooling education for the digital economy.

While the vision of participatory and global learning that this well-financed "EdTech" sector has helped popularize would likely earn praise from the early AV advocates of yesteryear as a sign of significant educational and social progress, it has increasingly been criticized by scholars as an extension of American hegemony and "electronic colonization."[29] Of particular concern are the US-led technological projects designed with the aim of improving education in low-income communities in the United States and global South. These include the One Laptop per Child project, an effort by researchers at the MIT Media Lab to provide low-income children with affordable computers, and Encarta Africana, Microsoft's multimedia encyclopedia program on global Black history and culture. These programs, while lauded by some observers for empowering underserved communities in the United States and abroad with critical digital and information literacies, have also raised concerns that such technology can be "another form of cultural imperialism used by corporate-led Western powers to force feed western values and worldviews to the rest of the world."[30]

Meanwhile, in the United States, the rhetoric of global citizenship, connectivity, and mutual understanding continues to be mobilized by a booming for-profit tech and start-up industry to legitimize its products for popular and educational consumption, and burnish its image as a socially responsible provider of a public good. Not unlike the stereograph, lantern slide, motion picture, and radio industries in the first decades of the twentieth century, the corporate leaders of today's (digital) media landscape have been quick to recognize educational partnerships, at home and abroad, as a form of free advertising and space in which to earn the public's "goodwill."[31] Today, as a century ago, schools are a powerful gateway for demonstrating to the general public the social benefits of new and controversial devices. And promotional images of far-flung racial and ethnic others using and benefiting from technology for instructional purposes continue to serve as a potent symbol of new products' universal appeal and global reach.

Figure 6.2
The One Laptop per Child project highlights how "joyful collaboration in Kenya" is facilitated with the help of affordable computers. One Laptop per Child, http://one .laptop.org/.

"Making the World More Open and Connected" and "Bringing the World Closer Together" are among the slogans that Facebook, the social media behemoth founded by Mark Zuckerberg in 2004, has adopted in recent years to foreground the company's positive contributions to society amid a swirl of controversies about data privacy.[32] Likewise, Facebook and other Silicon Valley giants have launched an array of digital services to get their products into schools.[33] Many of these draw on the century-old utopian claims of new media as a platform for virtual travel, intercultural exchange, and skill building for a global age. With the Google Expeditions app, for example, students can use a smartphone or mobile device, outfitted with a Google Cardboard virtual reality viewer, "to virtually explore an art gallery or museum, swim underwater, or navigate outer space, without leaving the classroom." Skype, the popular video chat platform owned by Microsoft, attracts educators with "virtual field trips" in which students can communicate online with faraway scientists and explorers, and a "global guessing game" called "Mystery Skype" in which two classrooms located in different parts of the world connect online and try to guess each other's location through questions about geography, language, and culture.[34]

There is no doubt that these initiatives can enhance students' learning in school, ignite their interests in different cultures, histories, and geographies, and promote mutual understanding by facilitating easier

Figure 6.3
A student in Eagle Grove, Iowa, virtually visits Dubai using Google Cardboard and the Google Expeditions app. "Bring Your Lessons to Life with Expeditions," Google, 2018, https://www.google.com/expeditions/.

communication across borders. The problem is when claims about technology's ability to democratize a world of knowledge for young people are taken at face value, or unmoored from larger, needed discussions about the long-standing power imbalances that have shaped global communication and media-enhanced learning.[35] The disconnect between technology's perpetual promise to create a better world and the persistence of intercultural inequalities in schools and society has contributed to the growth of several streams of critical research that aim to unpack the biases as well as assumptions implicit in the rhetoric of the "global village."[36] I'll conclude by mentioning a few of these, which constitute promising paths forward in the quest to incorporate technology meaningfully into education, and promote an intercultural understanding that is more equitable, reflexive, and just.

New Directions

Historicizing Emerging Technologies

First, there has been a long-building interest in both the humanities and social sciences to think critically as well as historically about the technologies we use to learn about and communicate with the world. This begins with shedding the "new era thinking" that is so common in moments of technological change in favor of situating new innovations within larger social, economic, and historical contexts.[37] There is a tendency in present-day promotions of educational technology to not only show a kind of ahistorical "amnesia" toward earlier technological experiments but also frame complex social and educational problems in terms that new devices can readily "fix."[38] The drive to cast new information technologies as social panaceas has been so consistent since the early twentieth century that, as Bettina Fabos has put it, "predictions for one educational technology can easily be substituted for another."[39] This ahistoricism has hindered a more robust and realistic debate about the potential contributions and limitations that particular devices bring to education. It has also diverted attention away from the stubborn structural and social problems—such as inadequate school facilities, underpaid teachers, widening socioeconomic inequality, and implicit biases against differences of race, ethnicity, class, ability, and gender—that demand deeper and more complex solutions than any single educational machine, device, or initiative can provide.[40]

Technology and Intercultural Understanding

Second, researchers have begun to trouble the assumed link between increased technology use and improved intercultural awareness, empathy, and understanding. As mentioned above, one concern is that the utopian "global village" rhetoric perpetuates Western hegemonic thinking and policies, and hinders critical reflection on the imperialist origins and infrastructures of international information flows.[41] While the American technology sector claims to produce "universal" products with global appeal, its leadership remains predominantly white, Western, and male. Much as the early twentieth-century advertisements for audio and visual educational technologies figured white, middle-class users as the norm, today's digital landscape is similarly engineered to reflect the tastes, habits, preferences, and ideologies of this demographic.[42] As Charles Ess has argued, because computer-mediated technologies "embed Western cultural values and communicative preferences, this means that well-meaning efforts to 'wire the world' in the name of an ostensibly universal/cosmopolitan vision of electronic democracy, paradoxically enough, emerge as a form of 'computer-mediated colonization.'"[43] Similarly, educational researchers Michalinos

Zembylas and Charalambos Vrasidas critique the "global village narrative" as a "modernist myth that presents cyberculture as culturally neutral and equally approachable by all peoples," while contending that "on the contrary, such a narrative, by erasing cultural differences and national boundaries, can be seen as a form of colonialism."[44]

Far from being equally approachable by all, educational technologies continue to be accessed and used differently by young people depending on the financial and educational resources at their disposal, their access to social capital, and their facility with English, the dominant language of the internet and digital products. Often described as a "digital divide," this problem has increasingly come to be understood as not only a gap in "physical access to computers and connectivity but also access to the additional resources that allow people to use technology well."[45] Roderic Crooks has asserted that mere access to computational technology cannot be taken as helpful or empowering on its own, as many initiatives designed to increase technology access and use in communities of color have also extracted valuable capital from them in the form of attention, labor, personal data, and time.[46] It's apparent that while there is growing consensus in the educational field today—building on the ideas of visual educators before World War II—that media-enriched learning should be "active," "participatory," and anchored in critical and creative activity to be truly democratic, there is still a great deal of work to be done to ensure that these kinds of critical engagements with educational technology do not remain limited to the already advantaged.[47]

In addition to examining the imperialist origins as well as unequal uses and impacts of technology in different learning communities, scholars have begun to call attention to the pervasive presence and influence of digital disinformation campaigns, extremist propaganda, and hate groups that sow racism, sexism, misogyny, and distrust online. White supremacist, religious extremist, and other antidemocratic groups and ideologies, some of which have managed to build large followings on popular social media platforms such as Facebook, YouTube, and Twitter, complicate the industry's preferred narrative of a more "connected world" being one that inevitably fosters more democratic and prosocial forms of "human connection."[48] The election of Donald Trump to the US presidency in 2016 and rise of far-right politicians in Europe at the time of this writing further underscore the persistence of nativist, white supremacist, and xenophobic ideologies in the West, described by some as an "anti-globalist backlash" to an intricately interconnected world economy that has seen a significant rise in labor automation and outsourcing, migration, and income inequality in the last few decades.[49]

Together, these streams of research suggest that more critical attention is needed to understand how matters of identity, access and use, entrenched inequality, and historical power relations play out in processes of global learning and intercultural interaction in technology-mediated contexts. The previous chapters of this book have tried to show that merely equipping teachers and students with tools to beam the world into the classroom did not necessarily create a more democratic media landscape or even bring about a deeper international understanding. Rather, many sensory aids and activities designed to promote world-mindedness ended up reinforcing existing racial and ethnic stereotypes as well as promoting self-congratulatory narratives about American benevolence, exceptionalism, and harmonious multiculturalism, while sidestepping the real and wide-ranging problems of imperialism, white supremacy, and structural racism.

Reimagining (Global) Citizenship

A third direction of research considers the definitions of citizenship that young people are encouraged to develop through their encounters with educational media. In the period discussed in this book, reformers began talking about novel approaches to communication in the classroom as necessary for preparing children for informed citizenship in an increasingly mediated and internationalized world. Importantly, their discourses contained early rumblings of grassroots and critical approaches to media reception, production, and circulation. They urged children not to just passively consume mediated information, as they presumably would when reading a textbook or watching a film in a theater, but to actively create, collect, exchange, and evaluate multiple forms of mediated information. This approach to teaching would ostensibly mold young people into liberal, freethinking subjects capable of considering multiple points of view, resisting the "mechanical thinking" associated with unquestioning patriotism and fascism, and participating in the public sphere. Using media creatively in the classroom would, as Stella Center of the National Council of Teachers of English remarked in 1933, help teachers "counter-balance" the "effect of the machine on the human being" and "help children live enriched lives in

the Machine Age," "heighten the imagination, refine the sensibilities, educate the emotions, and develop the sympathies," and create "an electorate
capable of weighing speeches, newspapers, and magazines, and of holding steady in the swirling currents of conflicting opinion."[50] Forging this
new ideal of a broad-minded, media-literate citizen during the birth of the
broadcast era, reformers' took a significant stride toward the development
of critical media literacy and intercultural education in the United States.
But their utopian aims were hobbled by their exclusiveness to white schools
along with their inattention to, or conscious avoidance of, contentious
social and political problems, such as the exclusion of racial and ethnic
minorities from the mainstream educational system and media industries.

Identifying these historically overlooked areas can throw light on some
of the enduring challenges in today's efforts to teach civic and technological literacies in schools. Most important, scholars have noted that there
is a "civic opportunity gap" that parallels the persistent racial and socioeconomic segregation in the US educational system. Schools in affluent,
predominantly white communities continue to offer more opportunities to
develop students' "civic capacities and commitments" than schools in low-
income and predominantly minority areas.[51] Additionally, white identity
continues to be centered both in the teaching profession, where a dispro-
portionate number of teachers are white despite the growing racial and ethnic diversity of students, and in school curricula, inhibiting the educational
system from fulfilling its potential to prepare students for deeper intercultural understanding and democratic, equitable living.[52] Responding to this,
James Banks has called for challenging the "assimilationist, liberal, and universal" definition of citizenship that has dominated in Western political
and educational discourse since the early twentieth century. Instead of this
definition, which assumes that young people must cast aside or downplay
their nondominant racial, ethnic, or linguistic identities in order to fully
participate in the national civic culture, Banks calls for an expanded definition of citizenship that recognizes and centers different group identities
in the process of sustaining democracy and achieving structural equality.[53]

Correspondingly, there is a move to think more broadly and reflexively
about the definition of global citizenship that is taught in the United States
and other Western societies. The definition of world citizenship in American education has been, since its emergence in the early twentieth century,
persistently apolitical, structured around what Talya Zemach-Bersin calls a

rhetoric of "universal kinship and belonging" that obscures or glosses over the historical as well as contemporary policies than unequally distribute power and wealth by race, class, gender, and nationality.[54] Today, educators and researchers encourage a more politically engaged and critical perspective on global citizenship.[55] They warn against the assumption that the increasing globalization of culture, commerce, and politics will automatically yield more world-minded or tolerant subjectivities in young people. Without educating students on the social, political, and ethical dimensions of global interdependence, the processes of globalization can just as easily "give rise to profoundly conservative ethnocultural affiliations and largely instrumental notions of global citizenship" that emphasize the accumulation of cultural capital and economic competitiveness over qualities like mutuality, empathy, solidarity, and deep understanding.[56]

Finally, the rise of social media and participatory digital culture since the early 2000s has renewed a debate about what constitutes meaningful civic education and engagement in the public sphere. While some suggest that society could benefit more from teaching about "capital 'P' politics," voting, and the workings of government, others argue for greater recognition and support of young people's diverse cultural productions, "remixes," identity expressions, and self-representations online and off-line as legitimate forms of democratic action.[57] This "cultural citizenship" view evokes that of Dewey's. At the turn of the twentieth century, he believed that schools could play a central role in revitalizing democracy by more deeply engaging students in their communities and the arts. As Dewey remarked in 1902, "We find that most of our pressing political problems cannot be solved by special measures of legislation or executive activity; but only by the promotion of common sympathies and a common understanding."[58] Ideally, media in the classroom can be mobilized in ways that serve both these objectives, preparing students to mobilize diverse communication technologies to participate at once in shaping policy and creatively contributing to public culture. Educators can also utilize technology in unconventional ways to promote more critical reflections on oppressive systems and technologies themselves, through practices of "critical making" and projects designed around objectives of social justice.[59]

For over a century, American educationalists, technology producers, and government leaders have been prone to techno-idealistic assumptions and "new era thinking" about the transformative, uplifting impact that new technology will have on schools and society, both at home and abroad. The mind-set that media can bring the world into the classroom so as to efficiently mold students into more democratic and tolerant citizens has facilitated the entry of wave after wave of devices into schools, from turn-of-the-century stereographs, films, and *National Geographic* magazines, to interwar pen pal letters and pageants, to twenty-first-century computers and digital tablets. The enduring faith in these technologies of global citizenship has impeded a more critical discussion of how the "world picture" they render is shaped by political and commercial interests, historical and social forces, and enduring inequalities of power. Understanding how these tensions have shaped education over the last century is important for identifying a more equitable, and genuinely global and intercultural, path forward as media and technology become increasingly integral to education in the new millennium.

Notes

Chapter 1

1. Anonymous, "New Geography, Book II," *Visual Education* 1, no. 5 (1920): 41–42; James N. Emery, "Sources of Visual Aids at Moderate Cost," *Educational Screen* 3, no. 6 (1924): 211–213.

2. Lucille Greer, "Picture Potentialities in World Friendship," *Educational Screen* 3, no. 2 (1924): 49–50; Frank Morton McMurry, ed., *The World Visualized for the Classroom: 1000 Travel Studies through the Stereoscope and in Lantern Slides, Classified and Cross Referenced for 25 Different School Subjects* (New York: Underwood and Underwood, 1915); Stephen Petrina, "Getting a Purchase on 'The School of Tomorrow' and Its Constituent Commodities: Histories and Historiographies of Technologies," *History of Education Quarterly* 42, no. 1 (2002): 75–111.

3. Mizuko Ito, Kris Gutiérrez, Sonia Livingston, Bill Penuel, Jean Rhodes, Katie Salen, Juliet Schor, Julian Sefton-Green, and S. Craig Watkins, *Connected Learning: An Agenda for Research and Design* (Irvine, CA: Digital Media and Learning Research Hub, 2013).

4. John M. Carroll and Mary Beth Rosson, "The Neighborhood School in the Global Village," *IEEE Technology and Society Magazine* 17, no. 4 (1998): 4–9. For critiques of the "myth of the global village," see Gail E. Hawisher and Cynthia L. Selfe, eds., *Global Literacies and the World Wide Web* (New York: Routledge, 2005), 1–9, and Ramesh Srinivasan, *Whose Global Village?: Rethinking How Technology Shapes Our World* (New York: NYU Press, 2017); 6.

5. Charles R. Acland, "Never Too Cool for School," *Journal of Visual Culture* 13, no. 1 (2014): 13–16.

6. Katharyne Mitchell, "Geographies of Identity: The Intimate Cosmopolitan," *Progress in Human Geography* 31, no. 5 (2007): 709.

7. Joel H. Spring, *Images of American Life: A History of Ideological Management in Schools, Movies, Radio, and Television* (Albany: SUNY Press, 1992), 2.

8. Ronald Walter Greene, "Pastoral Exhibition: The YMCA Motion Picture Bureau and the Transition to 16mm, 1928–1939," in *Useful Cinema*, ed. Charles R. Acland and Haidee Wasson (Durham, NC: Duke University Press, 2011), 206.

9. David Edgerton, *The Shock of the Old: Technology and Global History since 1900* (London: Oxford University Press, 2007), xii–xiv.

10. Cited in Todd Oppenheimer, *The Flickering Mind: Saving Education from the False Promise of Technology* (New York: Random House, 2007), 3. Similar claims from Edison are documented in Gregory Mason, "Teaching by the Movies," *The Outlook*, August 22, 1914: 963–970.

11. Laurie Segall and Erica Fink, "Bill Gates' Classroom of the Future," *CNN.Com*, March 8, 2013, https://money.cnn.com/2013/03/08/technology/innovation/bill-gates -education/index.html. A 2017 investigative series in the *New York Times*, "Education Disrupted," chronicled the efforts by powerful Silicon Valley corporations, such as Facebook, Google, and Microsoft, to transform public education in ways that would promote use and sales of their products, capture students' data, and prepare students for participation in the digital economy. See Natasha Singer, "The Silicon Valley Billionaires Remaking America's Schools," *New York Times*, June 6, 2017, https://www .nytimes.com/2017/06/06/technology/tech-billionaires-education-zuckerberg-face book-hastings.html. See also Christo Sims, *Disruptive Fixation: School Reform and the Pitfalls of Techno-Idealism* (Princeton, NJ: Princeton University Press, 2017).

12. Robert C. Snider, "The Machine in the Classroom," *Phi Delta Kappan*, December 1992, 316–323; Seymour Papert, *The Children's Machine: Rethinking School in the Age of the Computer* (New York: Basic Books, 1993).

13. Larry Cuban, *Teachers and Machines: The Use of Classroom Technology since 1920* (New York: Teachers College Press, 1986); Larry Cuban, *Oversold and Underused: Computers in the Classroom* (Cambridge, MA: Harvard University Press, 2003).

14. On recovering the roles of women in the history of technology, see Jennifer S. Light, "When Computers Were Women," *Technology and Culture*, no. 3 (1999): 455–483; Marie Hicks, *Programmed Inequality: How Britain Discarded Women Technologists and Lost Its Edge in Computing* (Cambridge, MA: MIT Press, 2017); Marsha Gordon, "Nontheatrical Media," *Feminist Media Histories* 4, no. 2 (2018): 128–134. On student-made media and other texts, see Catherine Sloan, "'Periodicals of an Objectionable Character': Peers and Periodicals at Croydon Friends' School, 1826– 1875," *Victorian Periodicals Review* 50, no. 4 (2017): 769–786; Christine Woyshner, "'I Feel I Am Really Pleading the Cause of My Own People': US Southern White Students' Study of African-American History and Culture in the 1930s through Art and the Senses," *History of Education* 47, no. 2 (March 4, 2018): 190–208.

15. Estrid Sørensen, *The Materiality of Learning: Technology and Knowledge in Educational Practice* (New York: Cambridge University Press, 2009); Juha Herkman, Taisto Hujanen, and Paavo Oinonen, eds., *Intermediality and Media Change* (Tampere,

Finland: University of Tampere, 2012); Ladislaus Semali and Ann Watts Pailliotet, eds., *Intermediality: The Teachers' Handbook of Critical Media Literacy* (New York: Routledge, 2018); Gabriele Balbi and Paolo Magaudda, *A History of Digital Media: An Intermedia and Global Perspective* (New York: Routledge, 2018).

16. On the distinctively "integrated, cross-media" approach of Edgar Dale's work, see Charles R. Acland, "American AV: Edgar Dale and the Information Age Classroom," *Technology and Culture* 58, no. 2 (2017): 392–442. Fred Turner traces how intellectuals and artists in the postwar United States, some engaging with the philosophies of progressive educators, developed multimedia technologies, installations, and performances to promote ideals of liberalism, tolerance, and a "democratic self" that would stand in contrast to the conformist and repressive ideologies associated with communism and fascism, in *The Democratic Surround: Multimedia and American Liberalism from World War II to the Psychedelic Sixties* (Chicago: University of Chicago Press, 2013). Jason Palmeri notes how a "multimodal turn" in writing and composition pedagogy, commonly hailed as a product of the digital era, was evident as early as the 1960s in *Remixing Composition: A History of Multimodal Writing Pedagogy* (Carbondale: Southern Illinois University Press, 2012).

17. Roger Daniels, *Guarding the Golden Door: American Immigration Policy and Immigrants since 1882* (New York: Hill and Wang, 2004), 30.

18. John Dewey, "The School as Social Center," *The Elementary School Teacher* 3, no. 2 (1902): 78; Joe William Trotter, Jr., ed., *The Great Migration in Historical Perspective: New Dimensions of Race, Class, Gender* (Bloomington: Indiana University Press, 1991).

19. John Higham, *Strangers in the Land: Patterns of American Nativism, 1860–1965* (New Brunswick, NJ: Rutgers University Press, 2011).

20. Jeffrey Mirel, *Patriotic Pluralism: Americanization Education and European Immigrants* (Cambridge, MA: Harvard University Press, 2010).

21. Kristin L. Hoganson, *Consumers' Imperium: The Global Production of American Domesticity, 1865–1920* (Chapel Hill: University of North Carolina Press, 2010), 8; Brendan Goff, "The Heartland Abroad: The Rotary Club's Mission of Civic Internationalism" (PhD diss., University of Michigan, 2008).

22. Richard Popp, "Machine-Age Communication: Media, Transportation, and Contact in Interwar America," *Technology and Culture* 52, no. 3 (2011): 467; Stella S. Center, "The Responsibility of Teachers of English in Contemporary American Life," *English Journal* 22, no. 2 (1933): 98.

23. Julia F. Brown, "Projects in International Understanding and World-Peace," *English Journal* 23, no. 8 (1934): 649–650.

24. Don W. Campbell and G. F. Stover, "Teaching International-Mindedness in the Social Studies," *Journal of Educational Sociology* 7 (1933): 244–248; Brian D. Behnken

and Gregory D. Smithers, *Racism in American Popular Media: From Aunt Jemima to the Frito Bandito* (Santa Barbara, CA: Praeger, 2015).

25. Ella Adeline Busch, "How We Learn to Understand Our Neighbors," *Hispania* 6 (1923): 205–213; Edna A. Collamore, "The Why of Geography Exhibits," *Journal of Geography* 27 (1928): 152–156; Adeline Mair, "Quelling a 'Class of Babel' by a Unit on Foreign Cultures," *Clearing House* 13 (1939): 265–266.

26. I follow Kimberlé Crenshaw in using "Black" and "African American" interchangeably throughout this book and capitalizing "Black" because "Blacks, like Asians, Latinos, and other 'minorities,' constitute a specific cultural group and, as such, require denotation as a proper noun." Correspondingly, "white(s)" is not capitalized here "since whites do not constitute a specific cultural group" and thus do not require denotation as a proper noun. See footnote 6 in Kimberlé Crenshaw, "Mapping the Margins: Intersectionality, Identity Politics, and Violence against Women of Color," *Stanford Law Review* 43, no. 6 (1991): 1244.

27. Davison Douglas, *Jim Crow Moves North: The Battle over Northern School Segregation, 1865–1954* (Cambridge: Cambridge University Press, 2005); Joel Spring, *The American School: From the Puritans to the Trump Era* (New York: Routledge, 2018).

28. Pauline R. Powers, "Ah! Geography!," *Journal of Geography* 37 (1938): 275.

29. Dwayne Roy Winseck and Robert M. Pike, *Communication and Empire: Media, Markets, and Globalization, 1860–1930* (Durham, NC: Duke University Press, 2007); James Schwoch, *Global TV: New Media and the Cold War, 1946–69* (Urbana-Champaign: University of Illinois Press, 2009); Daya Kishan Thussu, *International Communication: Continuity and Change*, 2nd ed. (London: Hodder Education, 2006).

30. Hoganson, *Consumers' Imperium*, 5; Jeremy Tunstall, *The Media Are American: Anglo-American Media in the World* (London: Constable, 1977).

31. Important exceptions include Catherine A. Lutz and Jane L. Collins, *Reading National Geographic* (Chicago: University of Chicago Press, 1993); Stephanie L. Hawkins, *American Iconographic: National Geographic, Global Culture, and the Visual Imagination* (Charlottesville: University of Virginia Press, 2010); Jani L. Barker, "'A Really Big Theme': Americanization and World Peace—Internationalism and/as Nationalism in Lucy Fitch Perkins's Twins Series," in *Internationalism in Children's Series* (London: Palgrave Macmillan, 2014), 76–94; Meenasarani Linde Murugan, "Exotic Television: Technology, Empire, and Entertaining Globalism" (PhD diss., Northwestern University, 2015).

32. David Morley, "For a Materialist, Non-Media-Centric Media Studies," *Television and New Media* 10, no. 1 (2008): 114–116; Arjun Appadurai, *The Social Life of Things: Commodities in a Cultural Perspective* (Cambridge: Cambridge University Press, 1997).

33. David Morley, "On Living in a Techno-Globalised World: Questions of History and Geography," *Telematics and Informatics* 30, no. 2 (2013): 61–62.

34. Larry Cuban, *How Teachers Taught: Constancy and Change in American Classrooms, 1890–1990* (New York: Teachers College Press, 1993).

35. Brian Goldfarb, *Visual Pedagogy: Media Cultures in and beyond the Classroom* (Durham, NC: Duke University Press, 2002); Allison Perlman, *Public Interests: Media Advocacy and Struggles over U.S. Television* (New Brunswick, NJ: Rutgers University Press, 2016); Victoria Cain, "From *Sesame Street* to Prime Time School Television: Educational Media in the Wake of the Coleman Report," *History of Education Quarterly* 57, no. 4 (November 2017): 590–601. For further reading on the history of educational and nontheatrical film, see Charles R. Acland and Haidee Wasson, eds., *Useful Cinema* (Durham, NC: Duke University Press, 2011); Devin Orgeron, Marsha Orgeron, and Dan Streible, eds., *Learning with the Lights Off: Educational Film in the United States* (New York: Oxford University Press, 2012).

36. Spring, *Images of American Life*; Stuart J. Foster, "Pride and Prejudice: Treatment of Immigrant Groups in United States History Textbooks, 1890–1930," *Education and Culture* 17, no. 1 (2001): 1–7; Michael Apple and Linda Christian-Smith, *The Politics of the Textbook* (New York: Routledge, 1991); Orgeron, Orgeron, and Streible, *Learning with the Lights Off*; Bill Bigelow, "On the Road to Cultural Bias: A Critique of *The Oregon Trail* CD-ROM," *Language Arts* 74, no. 2 (1997): 84–93; Katharine Slater, "Who Gets to Die of Dysentery?: Ideology, Geography, and *The Oregon Trail*," *Children's Literature Association Quarterly* 42, no. 4 (November 10, 2017): 374–395.

37. Victoria E. M. Cain, "Seeing the World: Media and Vision in US Geography Classrooms, 1890–1930," *Early Popular Visual Culture* 13, no. 4 (2015): 279–280; Susan Schulten, *The Geographical Imagination in America, 1880–1950* (Chicago: University of Chicago Press, 2001).

38. E. Winifred Crawford, "The Teacher's and Pupil's Use for Visual Aids in Studying Geography," *Visual Education* 2, no. 6 (1921): 6–11.

39. Ercel C. McAteer, "The Influence of Motion Pictures upon the Development of International Co-Operation," *Educational Screen* 7, no. 3 (1928): 94; Ruth M. Whitfield, "Screen Travels—Courses in World Understanding," *Visual Education* 3, no. 7 (1922): 317.

40. Anne Friedberg, *Window Shopping: Cinema and the Postmodern* (Berkeley: University of California Press, 1993); Anne Friedberg, *The Virtual Window: From Alberti to Microsoft* (Cambridge, MA: MIT Press, 2006); Barbara Maria Stafford and Frances Terpak, *Devices of Wonder: From the World in a Box to Images on a Screen* (Los Angeles: Getty Publications, 2001); Barbara Maria Stafford, *Artful Science: Enlightenment Entertainment and the Eclipse of Visual Education* (Cambridge, MA: MIT Press, 1994).

41. Stafford and Terpak, *Devices of Wonder*, 6–7.

42. Dagmar Capkova, "J. A. Comenius's 'Orbis Pictus' in Its Conception as a Textbook for the Universal Education of Children," *Paedagogica Historica* 10, no. 1

(1970): 5–27; John Willinsky, *Learning to Divide the World: Education at Empire's End* (Minneapolis: University of Minnesota Press, 1998), 57; Sarah Anne Carter, *Object Lessons: How Nineteenth-Century Americans Learned to Make Sense of the Material World* (New York: Oxford University Press, 2018); Barker, "Really Big Theme"; McAteer, "Influence of Motion Pictures," 95, 101; Little Passports, 2019, https://www.little passports.com/.

43. Carolyn Marvin, *When Old Technologies Were New: Thinking about Electric Communication in the Late Nineteenth Century* (New York: Oxford University Press, 1990); Lynn Spigel, *Make Room for TV: Television and the Family Ideal in Postwar America* (Chicago: University of Chicago Press, 1992); X. Theodore Barber, "The Roots of Travel Cinema: John L. Stoddard, E. Burton Holmes and the Nineteenth-Century Illustrated Travel Lecture," *Film History* 5 (1993): 68–84; Lisa Parks, "Kinetic Screens: Epistemologies of Movement at the Interface," in *Media Space: Place, Scale and Culture in a Media Age*, ed. Nick Couldry and Anna McCarthy (New York: Routledge, 2004); Jeffrey Ruoff, ed., *Virtual Voyages: Cinema and Travel* (Durham, NC: Duke University Press, 2006).

44. Carl G. Rathmann, "Educational Museum of the St. Louis Public Schools," *US Bureau of Education Bulletin* 48, no. 622 (1914): 10–11.

45. Carl G. Rathmann, "The Value and Importance of the School Museum," *Journal of the National Education Association* 1, no. 1 (1916): 737.

46. On the history of white appropriations of Indian dress and imagery, see Philip Joseph Deloria, *Playing Indian* (New Haven, CT: Yale University Press, 1998).

47. Rathmann, "Educational Museum of the St. Louis Public Schools," 5.

48. William George Carr, *Education for World-Citizenship* (Palo Alto, CA: Stanford University Press, 1928), 9, 16.

49. McAteer, "Influence of Motion Pictures"; Center, "Responsibility of Teachers of English"; William Heard Kilpatrick, *Education for a Changing Civilization: Three Lectures Delivered on the Luther Laflin Kellogg Foundation at Rutgers University, 1926* (New York: Macmillan, 1932).

50. Elizabeth Miller Lobingier and John Leslie Lobingier, *Educating for Peace* (Boston: Pilgrim Press, 1930), 122.

51. Carr, *Education for World-Citizenship*, 9; E. Estelle Downing, "What English Teachers Can Do to Promote World-Peace," *English Journal* 14, no. 3 (1925): 190.

52. Lester K. Ade, "Can Education Stimulate an Effective Universal Social Participation for Modern Citizenship?," *Public Education Bulletin of Pennsylvania* 5, no. 2 (1938): 24; Hazel R. Smith, "Education for World-Citizenship," *School Magazine* 4, no. 8 (1922): 219–220.

53. Zoe Burkholder, *Color in the Classroom: How American Schools Taught Race, 1900–1954* (New York: Oxford University Press, 2011), 34–35. See also LaGarrett J. King, Christopher Davis, and Anthony L. Brown, "African American History, Race, and Textbooks: An Examination of the Works of Harold O. Rugg and Carter G. Woodson," *Journal of Social Studies Research* 36, no. 4 (2012): 359–386.

54. Ade, "Can Education Stimulate an Effective Universal Social Participation for Modern Citizenship?"

55. As an example of a technology of citizenship, Cruikshank describes the Juvenile Street Cleaning League of New York City, formed in the 1890s by the city's street-cleaning commissioner, to instill in working-class and immigrant children a feeling of responsibility for the disposal of litter and waste. Forming voluntary associations of children to monitor and dispose of waste in their neighborhoods, such measures were touted as teaching "civic pride" and empowering individuals to collectively solve social problems outside of government, while also encouraging "people to carry out the purpose of government" and "making the people self-governing." This was, according to Cruikshank, a "classically liberal and democratic technology to get the people to police themselves so that the street-cleaning commissioner did not have to." In *The Will to Empower: Democratic Citizens and Other Subjects* (Ithaca, NY: Cornell University Press, 1999), 8-10.

56. Mary Burnett Talbert, "The Negro's Right to World Citizenship," in *Addresses at the Third World's Christian Citizenship Conference* (Pittsburgh: National Reform Association, 1919), 262.

57. Bruce Robbins, *Feeling Global: Internationalism in Distress* (New York: NYU Press, 1999), 18, cited in Talya Zemach-Bersin, "Global Citizenship and Study Abroad: It's All about U.S.," *Critical Literacy: Theories and Practices* 1 (2007): 20.

58. Zemach-Bersin, "Global Citizenship and Study Abroad"; Paul A. Kramer, "Is the World Our Campus? International Students and U.S. Global Power in the Long Twentieth Century," *Diplomatic History* 33, no. 5 (2009): 775–806; Vanessa de Oliveira Andreotti and Lynn Mario T. M. de Souza, eds., *Postcolonial Perspectives on Global Citizenship Education* (New York: Routledge, 2014).

59. Cruikshank, *Will to Empower*, 5.

60. John Dewey, "Nationalizing Education," *Journal of the National Education Association* 1, no. 1 (1916): 183–184.

Chapter 2

1. Frank Morton McMurry, ed., *The World Visualized for the Classroom: 1000 Travel Studies through the Stereoscope and in Lantern Slides, Classified and Cross Referenced for 25 Different School Subjects* (New York: Underwood and Underwood, 1915), iv.

For a biography of Dorris, see Wendell G. Johnson, "'Making Learning Easy and Enjoyable': Anna Verona Dorris and the Visual Instruction Movement, 1918–1928," *TechTrends: Linking Research and Practice to Improve Learning* 52, no. 4 (July 2008): 51–58. On the history of the visual education movement, see Paul Saettler, *The Evolution of American Educational Technology* (Greenwich, CT: Information Age Publishing, 2004), 123–176; Larry Cuban, *Teachers and Machines: The Use of Classroom Technology since 1920* (New York: Teachers College Press, 1986), 9–26; Devin Orgeron, Marsha Orgeron, and Dan Streible, eds., *Learning with the Lights Off: Educational Film in the United States* (New York: Oxford University Press, 2012); Elizabeth Wiatr, "Between Word, Image, and the Machine: Visual Education and Films of Industrial Process," *Historical Journal of Film, Radio and Television* 22, no. 3 (2002): 333–351; Victoria E. M. Cain, "'The Direct Medium of the Vision': Visual Education, Virtual Witnessing and the Prehistoric Past at the American Museum of Natural History, 1890–1923," *Journal of Visual Culture* 9, no. 3 (December 1, 2010): 284–303; Victoria E. M. Cain, "Seeing the World: Media and Vision in US Geography Classrooms, 1890–1930," *Early Popular Visual Culture* 13, no. 4 (2015): 276–292; Katie Day Good, "Making Do with Media: Teachers, Technology, and Tactics of Media Use in American Classrooms, 1919–1946," *Communication and Critical/Cultural Studies* 13, no. 1 (2016): 75–92.

2. Anna Verona Dorris, "Educating the Twentieth-Century Youth," *Junior-Senior High School Clearing House* 5, no. 4 (1930): 203.

3. Saettler, *Evolution of American Educational Technology*, 57–64; William J. Reese, "The Origins of Progressive Education," *History of Education Quarterly* 41, no. 1 (2001): 1–24; Cuban, *Teachers and Machines*, 10–11.

4. W. E. Graves, "Imagery, Thinking, Visualization," *Educational Screen* 6, no. 6 (1927): 261–262, 299; G. H. Bretnall, "A Neglected Phase of Visual Education," *Educational Screen* 10, no. 4 (1931): 107.

5. On late nineteenth- and early twentieth-century conceptualizations of communication technology and its social impact, see Daniel Czitrom, *Media and the American Mind: From Morse to McLuhan* (Chapel Hill: University of North Carolina Press, 1982); John Durham Peters, "Satan and Savior: Mass Communication in Progressive Thought," *Critical Studies in Mass Communication* 6 (1989): 247–263; Richard Popp, "Machine-Age Communication: Media, Transportation, and Contact in Interwar America," *Technology and Culture* 52, no. 3 (2011): 459–484; on armchair travel, Alison Byerly, *Are We There Yet?: Virtual Travel and Victorian Realism* (Ann Arbor: University of Michigan Press, 2012); Kristin L. Hoganson, *Consumers' Imperium: The Global Production of American Domesticity, 1865–1920* (Chapel Hill: University of North Carolina Press, 2010), 153–208.

6. On educational film as a "magic carpet," see Marie E. Goodenough, "Journey Geography by the Visual Route," *Educational Screen* 1, no. 7 (1922): 229. On radio as a "magic chain," see Benjamin Harrison Darrow, "Radio as an Educational Agency"

(broadcast sponsored by the National Congress of Parents and Teachers, National Broadcasting Company, January 17, 1935), Ohio School of the Air, RG8.D.6, box 1, folder 3/59, Ohio State University. On the potential of educational television to "provide learners with a magic carpet to a wider world of experience," see William Van Til, "Is Progressive Education Obsolete?," *Saturday Review*, February 17, 1962. On the promise of film and radio "bringing the world into the classroom," see Cuban, *Teachers and Machines*, 9–26.

7. Dan H. Wishnietsky, *Using Computer Technology to Create a Global Classroom* (Bloomington, IN: Phi Delta Kappa Educational Foundation, 1993), 15, 24; David Selby and Graham Pike, "Global Education: Relevant Learning for the Twenty-First Century," *Convergence* 33 (2000): 138–149; Elizabeth Langran and Irene Langran, "Technology's Role in Global Citizenship Education," in *Globalization and Global Citizenship: Interdisciplinary Approaches*, ed. Irene Langran and Tammy Birk (London: Routledge, 2016), 56–68.

8. Jennifer Daryl Slack, "The Theory and Method of Articulation in Cultural Studies," in *Stuart Hall: Critical Dialogues in Cultural Studies*, ed. David Morley and Kuan-Hsing Chen (London: Routledge, 1996), 113–130.

9. Carolyn Marvin, *When Old Technologies Were New: Thinking about Electric Communication in the Late Nineteenth Century* (New York: Oxford University Press, 1990), 191–231; James W. Carey and John J. Quirk, "The Mythos of the Electronic Revolution," *American Scholar* 39, no. 3 (1970): 395–424.

10. William J. Reese, *Power and the Promise of School Reform: Grassroots Movements during the Progressive Era* (New York: Teachers College Press, 2002); Joel H. Spring, *Images of American Life: A History of Ideological Management in Schools, Movies, Radio, and Television* (Albany: SUNY Press, 1992); Thomas D. Snyder, "120 Years of American Education: A Statistical Portrait" (Washington, DC: Office of Educational Research and Improvement, US Department of Education, 1993).

11. John Dewey and Evelyn Dewey, *Schools of To-Morrow* (New York: E. P. Dutton and Company, 1915); John Dewey, *Democracy and Education: An Introduction to the Philosophy of Education* (New York: Macmillan, 1916).

12. H. C. Dollison, "Visual Instruction," *Normal Instructor and Primary Plans* 29, no. 4 (1920): 36, 62; W. A. Willson, "The Phonograph in Americanization," *Proceedings of the Americanization Conference*, 1919, 50–60; Jennifer Peterson, *Education in the School of Dreams: Travelogues and Early Nonfiction Film* (Durham, NC: Duke University Press, 2013), 104–116; Spring, *Images of American Life*, 54–56, 103–104; Stephanie L. Hawkins, *American Iconographic: National Geographic, Global Culture, and the Visual Imagination* (Charlottesville: University of Virginia Press, 2010), 30–41; Amanda R. Keeler, "John Collier, Thomas Edison and the Educational Promotion of Moving Pictures," in *Beyond the Screen: Institutions, Networks, and Publics of Early Cinema*, ed.

Marta Braun, Charlie Keil, Rob King, Paul Moore, and Louis Pelletier (Bloomington: Indiana University Press, 2012), 117–125.

13. John Dewey, "Democracy in Education," *Elementary School Teacher* 4, no. 4 (1903): 200–204.

14. Underwood and Underwood, "One of the Greatest World Events," *School Board Journal* 58, no. 2 (1919): 82; George Kleine, *Catalogue of Educational Motion Picture Films* (Chicago: Bentley, Murray and Co., 1910). For the parallels between Kleine's catalog, Starr's essay, and earlier technologies of virtual travel and visual display, such as cabinets of curiosities, see Oliver Gaycken, *Devices of Curiosity: Early Cinema and Popular Science* (New York: Oxford University Press, 2015), 140–141.

15. Barbara Maria Stafford, *Artful Science: Enlightenment Entertainment and the Eclipse of Visual Education* (Cambridge, MA: MIT Press, 1994), 6–11, 97–102; Barbara Maria Stafford and Frances Terpak, *Devices of Wonder: From the World in a Box to Images on a Screen* (Los Angeles: Getty Publications, 2001), 6–7; Romita Ray, "The Beast in a Box: Playing with Empire in Early Nineteenth-Century Britain," *Visual Resources* 22, no. 1 (2006): 7–31.

16. Richard Daniel Altick, *The Shows of London* (Cambridge, MA: Harvard University Press, 1978), quoted in Anne Friedberg, *Window Shopping: Cinema and the Postmodern* (Berkeley: University of California Press, 1993), 24.

17. Tony Bennett, *The Birth of the Museum: History, Theory, Politics* (New York: Routledge, 1995); John Willinsky, *Learning to Divide the World: Education at Empire's End* (Minneapolis: University of Minnesota Press, 1998), 57.

18. X. Theodore Barber, "The Roots of Travel Cinema: John L. Stoddard, E. Burton Holmes and the Nineteenth-Century Illustrated Travel Lecture," *Film History* 5 (1993): 70.

19. Alison Griffiths, *Wondrous Difference: Cinema, Anthropology and Turn-of-the-Century Visual Culture* (New York: Columbia University Press, 2002); Brenton J. Malin, "Looking White and Middle-Class: Stereoscopic Imagery and Technology in the Early Twentieth-Century United States," *Quarterly Journal of Speech* 93, no. 4 (2007): 403–424.

20. For an account of how competing impulses toward nativism and pluralism influenced prewar education, particularly the efforts to Americanize newly arrived immigrants, see John Higham, *Strangers in the Land: Patterns of American Nativism, 1860–1965* (New Brunswick, NJ: Rutgers University Press, 2011). After World War I, some progressive educators eschewed overt Americanization instruction in favor of a more multiculturalist approach that celebrated "cultural gifts," or the contributions of immigrants and minorities to American culture. Diana Selig, *Americans All: The Cultural Gifts Movement* (Cambridge, MA: Harvard University Press, 2008). The tension between teaching nationalism and internationalism was apparent in American

school materials on geography and commerce, which presented a hierarchical view of the world's nations according to their industrial development and engagement in trade, and depicted the United States and Europe as the pinnacle of progress and center of civilization. Clif Stratton, *Education for Empire: American Schools, Race, and the Paths of Good Citizenship* (Oakland: University of California Press, 2016), 145–150; Susan Schulten, *The Geographical Imagination in America, 1880–1950* (Chicago: University of Chicago Press, 2001), 106–117.

21. After winning reelection on a progressive, antiwar platform in 1916, President Woodrow Wilson reversed course and called the nation to war in 1917. Striving to cultivate national unity in the war effort, he appealed to school officials to teach about the conflict as a battle for democratic values at home and abroad. This included teaching a "broadened conception of national life" that simultaneously emphasized local, national, and international cooperation, or "the close dependence of individual on individual and nation on nation." Woodrow Wilson, "President Wilson on Problems of Community and National Life in the Schools," *School and Society* 6, no. 145 (1917): 404; David Kennedy, *Over Here: The First World War and American Society* (New York: Oxford University Press, 2004), 45–59. On the notion that world citizenship and patriotism need not be "opposing forces," see William George Carr, *Education for World-Citizenship* (Palo Alto, CA: Stanford University Press, 1928), 9–10; Susan Zeiger, "Teaching Peace: Lessons from a Peace Studies Curriculum of the Progressive Era," *Peace and Change* 25, no. 1 (2000): 52–69.

22. Isaac L. Kandel, "International Understanding and the Schools," *NASSP Bulletin* 9, no. 1 (1925): 37; Ella Lyman Cabot, Fannie Fern Andrews, Fanny E. Coe, Mabel Hill, and Mary McSkimmon, *A Course in Citizenship and Patriotism* (Boston: Houghton Mifflin, 1918), xiii–xiv; E. Estelle Downing, "What English Teachers Can Do to Promote World-Peace," *English Journal* 14, no. 3 (1925): 183–192; E. Estelle Downing, "International Good Will through the Teaching of English," *English Journal* 14, no. 9 (1925): 675–685.

23. Selig, *Americans All*; Zeiger, "Teaching Peace."

24. Ruth M. Whitfield, "Screen Travels—Courses in World Understanding," *Visual Education* 3, no. 7 (1922): 315–317; James N. Emery, "Sources of Visual Aids at Moderate Cost," *Educational Screen* 3, no. 6 (1924): 211–213; Lucille Greer, "Picture Potentialities in Relation to World Peace," *Educational Screen* 8, no. 3 (1929): 71, 90; Dorris, "Educating the Twentieth-Century Youth"; Cain, "Seeing the World."

25. Lee Grieveson, "Visualizing Industrial Citizenship," in *Learning with the Lights Off: Educational Film in the United States*, ed. Marsha Orgeron, Devin Orgeron, and Dan Streible (New York: Oxford University Press, 2012), 107–123.

26. Anna Verona Dorris, *Visual Instruction in the Public Schools* (Boston, 1928), 39–40, 222–228; Stella Evelyn Myers, "A Visual Study of the Panama Canal," *Educational Screen* 7, no. 1 (1928): 30–32.

27. Ercel C. McAteer, "The Influence of Motion Pictures upon the Development of International Co-Operation," *Educational Screen* 7, no. 3 (1928): 94–95, 101; Ercel C. McAteer, "The Influence of Motion Pictures in Counteracting Un-Americanism," *Educational Screen* 7, no. 4 (1928): 140–141.

28. Malin, "Looking White and Middle-Class."

29. Tom Gunning, "'The Whole World within Reach': Travel Images without Borders," in *Virtual Voyages: Cinema and Travel*, ed. Jeffrey Ruoff (Durham, NC: Duke University Press, 2006), 25–41; Hoganson, *Consumers' Imperium*, 174–175; Griffiths, *Wondrous Difference*; Lauren Rabinovitz, "From Hale's Tours to Star Tours: Virtual Voyages, Travel Ride Films, and the Delirium of the Hyper-Real," in *Virtual Voyages: Cinema and Travel*, ed. Jeffrey Ruoff (Durham, NC: Duke University Press, 2006), 42–60.

30. Underwood and Underwood claimed that it invented the boxed tour format in the 1890s. Anonymous, *Manual of Instruction from Underwood and Underwood* (New York: Underwood and Underwood, 1900), 41–43. In addition to photographic tours, Underwood and Underwood produced more didactic products that it marketed simultaneously to consumers and educators, including a comprehensive "travel system" organized around particular countries and regions supplemented with ancillary texts, such as plotted maps and "guides" written by prominent scholars and travel experts. Lining its guidebooks with the glowing reviews of "educational patrons," such as military academies and university professors, the company hoped to showcase the edifying potentials of the virtual stereoscopic "tour" in both private leisure and public instructional settings. See James Ricalton, *India through the Stereoscope: A Journey through Hindustan* (New York: Underwood and Underwood, 1900).

31. Barber, "Roots of Travel Cinema"; Louis Walton Sipley, "The Magic Lantern," *Pennsylvania Arts and Sciences*, December 1939, 39–43.

32. "A Journey around the World in One Evening" was the title of a lantern show performed in Brooklyn in November 1865. Barber, "Roots of Travel Cinema," 69. See also Rick Altman, "From Lecturer's Prop to Industrial Product: The Early History of Travel Films," in *Virtual Voyages: Cinema and Travel*, ed. Jeffrey Ruoff (Durham, NC: Duke University Press, 2006), 61–78; Charles Musser, *The Emergence of Cinema: The American Screen to 1907* (Berkeley: University of California Press, 1994), 15–54.

33. In the late 1890s, both the Keystone View Company and Underwood and Underwood established in-house educational departments. Meredith A. Bak, "Democracy and Discipline: Object Lessons and the Stereoscope in American Education, 1870–1920," *Early Popular Visual Culture* 10, no. 2 (2012): 150.

34. Douglas Clay Ridgley, *Teachers' Guide for the Use of the "600 Set" of Keystone Stereographs and Lantern Slides for Visual Instruction* (Meadville, PA: Keystone View Company, 1911), 112.

35. Bak, "Democracy and Discipline"; Cain, "Seeing the World."

36. Mark S. W. Jefferson, "Stereoscopes in School," *Journal of Geography* 6 (1907): 154.

37. Keystone View Company, "Pictures Solve Many Classroom Difficulties," *Educational Screen* 3, no. 9 (1924): 363.

38. Frank Morton McMurry, "How to Study Stereographs and Lantern Slides," in *Visual Education through Stereographs and Lantern Slides* (Meadville, PA: Keystone View Company, 1917), ix; Dollison, "Visual Instruction."

39. McMurry, *World Visualized for the Classroom*, xiv; Ridgley, *Teachers' Guide*, 8.

40. James N. Emery, "The Slide Route to Africa," *Educational Screen* 5 (1926): 170; James N. Emery, "The Slide Route to India," *Moving Picture Age* 5, no. 8 (1922): 9–10, 26; James N. Emery, "South America via the Slide Route," *Moving Picture Age* 5, no. 2 (1922): 13–14, 35.

41. A. W. Abrams, "Collection, Organization, and Circulation of Visual Aids to Instruction by State Bureaus," *Journal of the National Education Association* 1 (1916): 743.

42. Griffiths, *Wondrous Difference*; Malin, "Looking White and Middle-Class," 409–410; Stuart Hall, "The Spectacle of the Other," in *Discourse Theory and Practice: A Reader*, ed. Simeon J. Yates, Stephanie Taylor, and Margaret Wetherell (Thousand Oaks, CA: SAGE, 2001), 324–344.

43. Mark Jefferson, "Racial Geography," in *Visual Education through Stereographs and Lantern Slides: School Work Visualized and Vitalized* (Meadville, PA: Keystone View Company, 1917), 30–45.

44. Selig, *Americans All*, 68–112.

45. Walter Lippmann, *Public Opinion* (New York: Harcourt Brace and Co., 1922); Albert Perry Brigham, "Geography and the War," *Journal of Geography* 19 (1920): 89–102.

46. Anonymous, "Schools Teaching Geography by Sight-Seeing Method," *School Education* 39, no. 2 (1919): 3.

47. Schulten, *Geographical Imagination in America*, 149–151.

48. Catherine Lutz and Jane Collins, "Becoming America's Lens on the World: National Geographic in the Twentieth Century," *South Atlantic Quarterly* 91, no. 1 (1992): 163, cited in Schulten, *Geographical Imagination in America*, 150.

49. Hawkins, *American Iconographic*, 117–118.

50. Jessie L. Burrall, "Visual Instruction in Geography," *Normal Instructor and Primary Plans*, October 1920, 70; Jessie L. Burrall, "Visualization," *American School* 6,

no. 7–8 (1920): 217; Jessie L. Burrall, "Sight-Seeing in School: Taking Twenty Million Children on a Picture Tour of the World," *National Geographic* 35 (1919): 497–500.

51. US Bureau of Education, "The School's Interest in World Geography," *School Life* 4, no. 9–10 (May 1, 1920): 9.

52. Anonymous, "Miscellaneous Notes," *Visual Education* 2, no. 4 (1921): 25.

53. Nelson L. Greene, "Foreword," *Visual Education* 1, no. 1 (1920): 4; Cline M. Koon, "The Use of Visual Equipment in Elementary and Secondary Schools," *Journal of the Society of Motion Picture Engineers* 28, no. 1 (1937); Good, "Making Do with Media"; Ellsworth C. Dent, *The Audio-Visual Handbook* (Chicago: Society for Visual Education, 1942), 12; Saettler, *Evolution of American Educational Technology*, 106–108.

54. Saettler, *Evolution of American Educational Technology*, 166–167.

55. Bettina Fabos, *Wrong Turn on the Information Superhighway: Education and the Commercialization of the Internet* (New York: Teachers College Press, 2004); Bettina Fabos and Michelle D. Young, "Telecommunication in the Classroom: Rhetoric versus Reality," *Review of Educational Research* 69, no. 3 (1999): 217–259.

56. Spring, *Images of American Life*; A. C. Derr, "Demand for Clean Films Develops Educational Pictures," *Reel and Slide*, March 1918, 7–8.

57. Lee Grieveson, *Policing Cinema: Movies and Censorship in Early-Twentieth-Century America* (Berkeley: University of California Press, 2004), quoted in Jennifer Peterson, "The Knowledge Which Comes in Pictures: Educational Films and Early Cinema Audiences," in *A Companion to Early Cinema*, ed. Andre Gaudreault, Nicolas Dulac, and Santiago Hidalgo (Hoboken, NJ: John Wiley and Sons, 2012), 280.

58. Colin N. Bennett, "Emotional Reactions to Educational Films," *Educational Film Magazine* 7, no. 2–3 (1922): 29; Richard Abel, "The 'Much Vexed Problem' of Nontheatrical Distribution in the Late 1910s," *Moving Image: The Journal of the Association of Moving Image Archivists* 16, no. 2 (2016): 91; Kleine, *Catalogue of Educational Motion Picture Films*; A. B. Jewett, "Millions Get Ford Message by 'Educational Weekly' Films," *Reel and Slide*, March 1918, 31.

59. Peterson, *Education in the School of Dreams*, 2, 104–116.

60. George Creel, *How We Advertised America* (New York: Harper and Brothers Publishers, 1920), 117–132; Martha Bayles, *Through a Screen Darkly: Popular Culture, Public Diplomacy, and America's Image Abroad* (New Haven, CT: Yale University Press, 2014), 124; K. Jack Bauer, *List of World War I Signal Corps Films* (Washington, DC: National Archives and Records Service, General Services Administration, 1957).

61. Bayles, *Through a Screen Darkly*, 124; Leslie Midkiff DeBauche, *Reel Patriotism: The Movies and World War I* (Madison: University of Wisconsin Press, 1997); Edgar Dale, "Motion Pictures and the War," *Educational Screen* 21, no. 6 (1942): 213–217.

62. Keeler, "John Collier, Thomas Edison and the Educational Promotion of Moving Pictures," 118; Saettler, *Evolution of American Educational Technology*, 100.

63. Thomas A. Edison, Inc., "In Classroom Work," *School Journal*, March 1914, 167.

64. Anonymous, "Motion-Picture Schoolhouses to Prevent Future Wars," *Current Opinion* 66, no. 4 (April 1919): 234; Keeler, "John Collier, Thomas Edison and the Educational Promotion of Moving Pictures."

65. Nelson L. Greene, "Motion Pictures in the Classroom," *Annals of the American Academy of Political and Social Science* 128 (1926): 122.

66. Phillip W. Stewart, "Henry Ford: Movie Mogul?," *Prologue* (Winter 2014); Mayfield Bray, *Guide to the Ford Film Collection in the National Archives* (Washington, DC: National Archives and Records Service, General Services Administration, 1970), 7; Anonymous, "How One State Handles Rural Movies," *Educational Film Magazine* 5, no. 5 (1921): 5; Beatrice Barrett, "Visual Education for Every School Everywhere," *Educational Film Magazine* 5, no. 2 (1921): 8; Beatrice Barrett, "Putting Visual Instruction within the Reach of Every School," *Moving Picture Age* (February 1921); "Ford Educational Library," *American Schoolmaster* 13, no. 7 (1920): 271–272.

67. Grieveson, "Visualizing Industrial Citizenship."

68. Jewett, "Millions Get Ford Message," 31. For examples of Ford's anti-Semitism, see Henry Ford and Samuel Crowther, *My Life and Work* (New York: Doubleday, Page and Company, 1923), 250–252. Ford authorized numerous articles promoting anti-Semitic theories of a global Jewish conspiracy in his newspaper, the *Dearborn Independent*, in the 1920s. For example, on the role of Jews in the movie industry, see Anonymous, "Jewish History in the United States," in *The International Jew: The World's Foremost Problem: A Reprint of a Series of Articles Appearing in the Dearborn Independent from May 22 to October 2, 1920* (Dearborn, MI: Dearborn Publishing Co., 1920), 1:39. Citing "the movies' moral failure," James Martin Miller, who corresponded with Ford and claimed to be his authorized biographer, wrote that Ford believed that Hollywood films "are not American and their producers are racially unqualified to reproduce the American atmosphere." In James Martin Miller and Henry Ford, *The Amazing Story of Henry Ford: The Ideal American and the World's Most Famous Private Citizen; a Complete and Authentic Account of His Life and Surpassing Achievements* (Chicago: M. A. Donohue and Company, 1922), 187.

69. Miller and Ford, *Amazing Story of Henry Ford*, 42–46; Stephen Meyer, "Adapting the Immigrant to the Line: Americanization in the Ford Factory, 1914–1921," *Journal of Social History* 14, no. 1 (1980): 67–82.

70. "Motion Photographs," *Sierra Educational News* 16, no. 1 (1920): 63.

71. Ford Motor Company, *New Orleans*, 1923, https://www.youtube.com/watch?v=1_AlYeOdP2A&t=333s; Ford Motor Company, *Los Angeles*, 1917, https://www

.youtube.com/watch?v=3Wp66mfUhT4&t=272s; Ford Motor Company, *Democracy in Education*, 1922, https://catalog.archives.gov/id/91176.

72. Will Hays, "Improvement of Moving Pictures," *Addresses and Proceedings of the National Education Association Annual Meeting*, 1922, 252, 257; Stephen Vaughn, "The Devil's Advocate: Will H. Hays and the Campaign to Make Movies Respectable," *Indiana Magazine of History* 101, no. 2 (2005).

73. Two prominent collaborations between the MPPDA and educators were the *Secrets of Success* and *Human Relations* film series in the 1930s. Saettler, *Evolution of American Educational Technology*, 107–108, 113; Craig Kridel, "Educational Film Projects of the 1930s: Secrets of Success and the Human Relations Films Series," in *Learning with the Lights Off: Educational Film in the United States*, ed. Devin Orgeron, Marsha Orgeron, and Dan Streible (New York: Oxford University Press, 2012), 215–229.

74. Spring, *Images of American Life*, 56–58; Saettler, *Evolution of American Educational Technology*, 145–147; Charles H. Judd, "Report of the Committee to Cooperate with the Motion Picture Producers," in *Addresses and Proceedings of the National Education Association* (Washington, DC: National Education Association, 1923). For examples of tech companies courting educators in the digital era, see Fabos, *Wrong Turn on the Information Superhighway*; Natasha Singer and Danielle Ivory, "How Silicon Valley Plans to Conquer the Classroom," *New York Times*, November 3, 2017, https://www.nytimes.com/2017/11/03/technology/silicon-valley-baltimore-schools.html.

75. Motion Picture Producers and Distributors of America, "Press Release," MPPDA Digital Archive, record no. 259, January 11, 1926, http://mppda.flinders.edu.au/records/259.

76. Will Hays, "Letter to Col. Jason S. Joy," October 18, 1922, MPPDA Digital Archive, record no. 13, http://mppda.flinders.edu.au/records/13; Jason S. Joy, "Motion Pictures and Peace," in *World Friendship: A Series of Articles Written by Some Teachers in the Los Angeles Schools and by a Few Others Who Are Likewise Interested in the Education of Youth*, ed. Evaline Dowling (Los Angeles: Committee on World Friendship, 1928), 32–34.

77. Joy, "Motion Pictures and Peace," 32–34.

78. Anonymous, "Nanook of the North: A True Story of the Arctic," *Visual Education* 3, no. 7 (1922): 327–330; Anonymous, "The Films in Review: Jungle Adventures," *Visual Education* 3, no. 6 (1922): 306–308. Recommendations for a variety of films can be found in the Films in Review section of *Visual Education* and the Theatrical Field section in the *Educational Screen*.

79. Herbert A. Jump, "The Religious Possibilities of the Motion Picture," *Film History* 14, no. 2 (2002): 222–223; Laurence R. Campbell, "The Screen and the Student," *Educational Screen* 7 (1928): 142–143; Cline M. Koon, "The Motion Picture in International Understanding," in *Motion Pictures in Education in the United*

States: A Report Compiled for the International Congress of Educational and Instructional Cinematography (Chicago: University of Chicago Press, 1934), 36–46; John Ellmore Hansen, "Why We Should Use Pictures in Teaching," *Junior-Senior High School Clearing House* 5, no. 4 (1930): 206–207; Goodenough, "Journey Geography by the Visual Route," 229.

80. Ercel C. McAteer, "The Educational Value of Motion Pictures," *Educational Screen* 8, no. 5 (1929): 137.

81. L. A. Wiley, "Wider Use of Pictures in Instruction," *Oregon Teachers Monthly* 22, no. 2 (1917): 77–80.

82. Whitfield, "Screen Travels," 315.

83. Trends in early educational motion picture use can be observed in comparing two national surveys of visual education, the first completed in 1923, and the second in 1936. The former is F. Dean McCluskey's unpublished *The Administration of Visual Education: A National Survey*, completed for the NEA Committee on Visual Education in cooperation with the MPPDA. Cited in Saettler, *Evolution of American Educational Technology*, 137–138. The latter is the National Visual Instruction survey, conducted in 1936 for the US Office of Education with a grant from the American Council on Education. Koon, "Use of Visual Equipment." See also Cuban, *Teachers and Machines*, 13–15.

84. Saettler, *Evolution of American Educational Technology*, 106.

85. Kridel, "Educational Film Projects of the 1930s"; Dent, *Audio-Visual Handbook*, 164.

86. Mark A. May, "Educational Possibilities of Motion Pictures," *Journal of Educational Sociology* 11, no. 3 (1937): 151.

87. These included the shifts from thirty-five- to sixteen-millimeter "safety" film stock, and from silent to sound film, both of which required different types of projection equipment and technical know-how from educators. Mal Lee and Arthur Winzenried, *The Use of Instructional Technology in Schools: Lessons to Be Learned* (Australia: Australian Council for Educational Research Ltd., 2009), 43–44.

88. Koon, "Use of Visual Equipment," 284–285.

89. Koon, "Motion Picture in International Understanding," 36.

90. Koon, "Use of Visual Equipment," 285.

91. Benjamin Harrison Darrow, *Radio: The Assistant Teacher* (Columbus, OH: R. G. Adams and Company, 1932), 79.

92. Robert McChesney, "The Payne Fund and Radio Broadcasting," in *Children and the Movies: Media Influence and the Payne Fund Controversy*, ed. Garth S. Jowett, Ian C.

Jarvie, and Kathryn H. Fuller (New York: Cambridge University Press, 1996), 303–335; Josh Shepperd, "Rockefeller Underwriting of Local, Regional, and National Educational Broadcasting Experiments, 1934–1940" (working paper, Rockefeller Archive Center Publications, 2013), 1–20.

93. Spring, *Images of American Life*; Amanda Keeler, "Defining a Medium: The Educational Aspirations for Early Radio," *Journal of Radio and Audio Media* 23, no. 2 (November 2016): 278–287; Cuban, *Teachers and Machines*, 19–26; Randall Patnode, "What These People Need Is Radio: New Technology, the Press, and Otherness in 1920s America," *Technology and Culture* 44 (2003).

94. Darrow, *Radio*, 231–237.

95. Benjamin Harrison Darrow, "A Brief History of Educational Broadcasting" (paper presented at the Ohio School of the Air Conference, November 22, 1929), Ohio School of the Air, RG8.D.6, box 1, folder 2/59, Ohio State University; Benjamin Harrison Darrow, *Radio Trailblazing: A Brief History of the Ohio School of the Air and Its Implications for Educational Broadcasting* (Columbus, OH: College Book Company, 1940), 18–31; William Bianchi, "The Wisconsin School of the Air: Success Story with Implications," *Educational Technology and Society* 5, no. 1 (2002): 141.

96. Anonymous, "California Children Like Lessons Taught by Radio," *New York Times*, March 1, 1925.

97. The school program was an iteration of another travelogue-inspired feature produced by the *Daily News* from 1924 through 1931 called the "Radio Photologues," in which travel lecturers gave talks over the airwaves on Saturday evenings while readers looked at selected images preprinted in the photogravure section of the paper.

98. Stella Evelyn Myers, "Eye and Ear Instruction," *Educational Screen* 6, no. 9 (1927): 435; Katie Day Good, "Listening to Pictures: Converging Media Histories and the Multimedia Newspaper," *Journalism Studies* 18, no. 6 (2017): 691–709; Katie Day Good, "Radio's Forgotten Visuals," *Journal of Radio and Audio Media* 23, no. 2 (November 2016): 364–368.

99. Darrow, "Radio as an Educational Agency.""

100. "Travelogs" (Ohio School of the Air Courier, 1936), Ohio School of the Air, RG8.D.6, box 1, folder 16/29, Ohio State University; Anonymous, "U.S. Office of Education Plans Radio Workshop Staffed by Relief Groups," *Broadcasting*, 1936, 22; Darrow, *Radio Trailblazing*, 72–73; J. Frank Beatty, "Uncle Sam on the Air, with Donated Time," *Broadcasting*, April 15, 1936, 11; Joy Hayes, "'Selling' America to Americans: New Deal Radio and Media Education," *Flow* (blog), May 19, 2015, https://www.flowjournal.org/2015/05/selling-america-to-americans/.

101. Charles G. Abbott, "Letter to Dr. C. M. Focken, University of Otago," March 25, 1941, SIA RU000083, Smithsonian Institution Editorial and Publications Division,

Records, 1847–1966, box 2, folder: Smithsonian Radio Program: Correspondence, 1936–1962, Smithsonian Institution; Michele Hilmes, *Network Nations: A Transnational History of British and American Broadcasting* (New York: Routledge, 2012), 124–125.

102. William Dow Boutwell, "The Radio World Is Yours," *Phi Delta Kappan* 21, no. 7 (1939): 345–347. Both Harold Ickes, US secretary of the interior, and John Studebaker, US commissioner of education, offered praise for *The World Is Yours*. "Statement Inserted in Transcript of Hearings at Congress, December 16, 1936," December 16, 1936, SIA RU000083, Smithsonian Institution Editorial and Publications Division, Records, 1847–1966, box 2, folder: National Conference on Educational Broadcasting, 1936–1937, Smithsonian Institution.

103. Radio scripts, SIA 05–124, Production Records, 1937–1942, 1980–1981, 2002–2003, boxes 1 and 2, Smithsonian Institution. Austin H. Clark, a Smithsonian scientist and advocate of educational radio programs, developed *The World Is Yours* both for schools and the general public. For more on Clark's ideas about how to make scientific radio talks entertaining, see Austin H. Clark Papers, Smithsonian Institution Archives; "Radio Talks," *Scientific Monthly* 35 (1932): 352–359. For more on the ubiquity of the "Old Time Naturalist" character in educational media, particularly nature films, see Jennifer Peterson, "Glimpses of Animal Life: Nature Films and the Emergence of Classroom Cinema," in *Learning with the Lights Off: Educational Film in the United States*, ed. Marsha Orgeron, Devin Orgeron, and Dan Streible (New York: Oxford University Press, 2012), 145–167.

104. Benjamin Harrison Darrow, "Radio Picks the Lock on the School Room Door," 1931, Ohio School of the Air, RG8.D.6, box 1, folder 3/59, Ohio State University.

105. Charles G. Abbott, "Letter to Mr. D. B. Murray," July 10, 1941, Record Unit 83, Editorial and Publications Division, Records, 1847–1966, box 2, folder: Smithsonian Radio Program: Correspondence, 1936–1962, Smithsonian Institution.

106. National Broadcasting Company, "Primitive Music," *The World Is Yours*, May 8, 1938, SIA 05–124, Production Records, 1937–1942, 1980–1981, 2002–2003, box 1, folder 3, Smithsonian Institution.

107. National Broadcasting Company, "Eskimos," *The World Is Yours*, n.d., SIA 05–124, Production Records, 1937–1942, 1980–1981, 2002–2003, box 1, folder 3, Smithsonian Institution.

108. "Statement Inserted in Transcript of Hearings at Congress, December 16, 1936."

109. Fan Correspondence, n.d., Record Unit 83, Editorial and Publications Division, Records, 1847–1966, box 2, folder: Smithsonian Radio Program: Correspondence, 1936–1962, Smithsonian Institution.

110. Saettler, *Evolution of American Educational Technology*, 214.

111. Saettler, *Evolution of American Educational Technology*, 114.

112. National Education Association, *Wartime Handbook for Education* (Washington, DC: National Education Association of the United States, 1943), 40–41; Charles R. Acland, "Curtains, Carts and the Mobile Screen," *Screen* 50, no. 1 (2009): 148–166.

113. "Post-War Growth Period, 1946–1957," Association for Educational Communications and Technology, 2001, https://aect.org/post-war_growth_period_1946-1.php.

114. Edgar Dale, *Audio-Visual Methods in Teaching* (New York: Dryden Press, 1946); Dent, *Audio-Visual Handbook*.

115. International Theatrical & Television Corporation, "These Children Must Be Readied for the World of Tomorrow," *Educational Screen* 24, no. 5 (1945): 201.

116. William Lewin, "Photoplays for International Understanding," *Educational Screen* 22, no. 10 (1943): 390.

Chapter 3

1. Edna A. Collamore, "The Why of Geography Exhibits," *Journal of Geography* 27 (1928): 154; Paul Saettler, *The Evolution of American Educational Technology* (Greenwich, CT: Information Age Publishing, 2004), 148.

2. Paul Mihailidis, *Media Literacy and the Emerging Citizen: Youth, Engagement and Participation in Digital Culture* (New York: Peter Lang, 2014); Renee Hobbs and Amy Jensen, "The Past, Present, and Future of Media Literacy Education," *Journal of Media Literacy Education* 1, no. 1 (2009), https://digitalcommons.uri.edu/jmle/vol1/iss1/1/.

3. Daniel C. Knowlton, "The Place of the Motion Picture in a Program of Visual Instruction," *Junior-Senior High School Clearing House* 5, no. 4 (1930): 195.

4. Charles R. Acland, "American AV: Edgar Dale and the Information Age Classroom," *Technology and Culture* 58, no. 2 (2017): 395. See also Brian Gregory, "Edgar Dale, Educational Radio, and Sensory Learning," *Antenna* (blog), March 16, 2015, http://blog.commarts.wisc.edu/2015/03/16/edgar-dale-educational-radio-and-sensory-learning/.

5. After the war, Dale was a member and later president of UNESCO's Commission on Technical Needs in Press, Film and Radio. In this and other advisory roles, he advocated for developing AV education as an agent of international understanding. Janet Leigh Hood-Hanchey, "The Transformation of Experience: An Historic Perspective on the Work of Edgar Dale" (PhD diss., University of Texas at Austin, 1981), 33; Edgar Dale, "A Proposal for International Audiovisual Exchange," *Audio Visual Communication Review* 8, no. 4 (1960): 229–233.

6. Acland, "American AV," 411–412.

7. Saettler, *Evolution of American Educational Technology*, 143–146, 163; "AECT in the 20th Century: A Brief History," Association for Educational Communications and Technology, 2001, https://aect.org/aect_in_the_20th_century_a_br.php.

8. M. Etienne, "The Opaque Projector," *American Biology Teacher* 4, no. 2 (1941): 62–63; Etta Schneider, "We Grow Up," *Educational Screen* 18, no. 1 (1939): 17.

9. A. W. Abrams, "Collection, Organization, and Circulation of Visual Aids to Instruction by State Bureaus," *Journal of the National Education Association* 1 (1916): 741.

10. Edward W. Stitt, "The Importance of Visual Instruction," *Journal of the National Education Association* 1 (1916): 737–738.

11. L. C. Everard, "Visual Material: Spur or Sedative," *Visual Education* 1, no. 5 (1920): 29; Charles W. Crumly, "The Movies: Bane or Blessing?," *Education* 40, no. 4 (1919): 199–213.

12. Philander P. Claxton, "Basic Material in Education," *Visual Education* 1, no. 6 (1920): 20–21; Anna Verona Dorris, *Visual Instruction in the Public Schools* (Boston, 1928), 37; Department of Public Instruction, Harrisburg, Pennsylvania, "Visual Education and the School Journey," *Educational Monographs* 1, no. 6 (1930): 5; Charles Francis Hoban, Charles Francis Hoban Jr., and Samuel B. Zisman, *Visualizing the Curriculum* (New York: Dryden Press, 1937), 30; Thomas Wendell, "The Stream of Perceptual Teaching," *Educational Screen* 18, no. 9 (1939): 326–327; Edgar Dale, *Audio-Visual Methods in Teaching* (New York: Dryden Press, 1946), 56–59.

13. Johann Amos Comenius and Charles Hoole, *Orbis Sensualium Pictus* (Sydney: Sydney University Press, 1967); Dagmar Capkova, "J. A. Comenius's 'Orbis Pictus' in Its Conception as a Textbook for the Universal Education of Children," *Paedagogica Historica* 10, no. 1 (1970): 5–27; Maria Esther Aguirre Lora, "Teaching through Texts and Pictures: A Contribution of Jan Amos Comenius to Education," *Revista Electrónica de Investigación Educativa* 3, no. 1 (2001), https://www.redib.org/recursos/Record/oai_articulo1179199-teaching-texts-pictures-contribution-jan-amos-comenius-education.

14. Sarah Anne Carter, "On an Object Lesson, or Don't Eat the Evidence," *Journal of the History of Childhood and Youth* 3, no. 1 (2010): 7–12.

15. Dorothy Parnell, "Picture Study in the Grades," *Elementary English Review* 6, no. 1 (1929): 24–27; Sally Gregory Kohlstedt, *Teaching Children Science: Hands-On Nature Study in North America, 1890–1930* (Chicago: University of Chicago Press, 2010).

16. Carter, "On an Object Lesson," 8; Sarah Anne Carter, *Object Lessons: How Nineteenth-Century Americans Learned to Make Sense of the Material World* (New York: Oxford University Press, 2018).

17. Katie Day Good, "Making Do with Media: Teachers, Technology, and Tactics of Media Use in American Classrooms, 1919–1946," *Communication and Critical/Cultural Studies* 13, no. 1 (2016): 75–92.

18. Claxton, "Basic Material in Education," 21; Crumly, "Movies," 204.

19. Anonymous, "Editorial," *Educational Screen* 5, no. 7 (1926): 389–390.

20. Joseph J. Weber, "Is the Term 'Visual Education' Scientific?," *Phi Delta Kappan* 11, no. 3 (1928): 79; Joseph J. Weber, "Comparative Effectiveness of Some Visual Aids in Seventh Grade Instruction" (PhD diss., Columbia University, 1922).

21. Harry Shaw, "Pocket and Pictorial Journalism," *North American Review* 243, no. 2 (1937): 299–300.

22. Dorris, *Visual Instruction in the Public Schools*, 36–38.

23. Hoban, Hoban, and Zisman, *Visualizing the Curriculum*, 260.

24. Saettler, *Evolution of American Educational Technology*, 140–143; Daniel C. Knowlton, "Equipping for Visual Education," *Junior-Senior High School Clearing House* 4 (1929): 198–202.

25. Dale, *Audio-Visual Methods in Teaching*, 37–52.

26. Acland, "American AV," 405; Edgar Dale, "Communicating with John Dewey," *News Letter* 25, no. 3 (December 1959).

27. Edgar Dale, "Coming to Our Senses," *News Letter: Bureau of Educational Research, Ohio State University* 4, no. 9 (1939): 1–2.

28. Dale, *Audio-Visual Methods in Teaching*, 38.

29. Collamore, "Why of Geography Exhibits," 156; N. L. Greene, "Motion Pictures in the Classroom," *Annals of the American Academy of Political and Social Science* 128 (1926): 122–130.

30. Charles Francis Hoban, "Possibilities of Visual-Sensory Aids in Education," *Educational Screen* 11 (1932): 198–200; Arnold P. Helfin, "Audio Aids in a Visual Program," *Educational Screen* 7, no. 2 (1938): 39–42; Wendell, "Stream of Perceptual Teaching"; G. Lester Anderson, "Should It Be Audio-Visual Aids or Audio-Visual Materials?," *Educational Screen* 24, no. 5 (1945): 198; A. C. Stenius, "Auditory and Visual Education," *Review of Educational Research* 15, no. 3 (1945): 243–255.

31. Saettler, *Evolution of American Educational Technology*, 166; F. Dean McClusky, "Pupil Constructed Science Exhibits," *Educational Screen* 8, no. 5 (1934): 136; David Goodman, *Radio's Civic Ambition: American Broadcasting and Democracy in the 1930s* (Oxford: Oxford University Press, 2011); Robert W. McChesney, *Telecommunications, Mass Media, and Democracy: The Battle for the Control of U.S. Broadcasting* (New York: Oxford University Press, 1993); James A. Findlay and Lillian Perricone, *WPA Museum*

Extension Project, 1935–1943: Government Created Visual Aids for Children from the Collections of the Bienes Museum of the Modern Book (Fort Lauderdale, FL: Bienes Museum of the Modern Book, 2009).

32. F. Dean McClusky, "Progressive Educators Meet," *Educational Screen* 7, no. 2 (1928): 79.

33. F. Dean McClusky, "Steady Growth," *Phi Delta Kappan* 22, no. 9 (1940): 412–413.

34. Helfin, "Audio Aids in a Visual Program," 39; Annette Glick, "Slide-Making and the Social Studies Laboratory, I," *Historical Outlook* 22, no. 5 (1931): 205–211; Dale, *Audio-Visual Methods in Teaching*, 38; Paula M. Kittel, "Pictures and Scrap Books as Aids in Teaching," *Monatshefte Für Deutschen Unterricht* 28, no. 3 (1936): 122. For more on teachers' aversion to mass-produced aids and their development of do-it-yourself classroom media, see Good, "Making Do with Media."

35. Ruth Bynum, "The Bulletin Board in English," *English Journal* 17, no. 3 (1928): 246–247; John Sterning, "Homemade Visual Aids," *See and Hear* 1, no. 8 (1946): 24.

36. Joseph J. Weber, "Aeroplane View of the Visual Aids Field, Part II," *Educational Screen* (1924): 338.

37. Ruth Messenger, "Assimilative Material," *English Journal* 23, no. 1 (1934): 58–63.

38. Garth S. Jowett, Ian C. Jarvie, and Kathryn H. Fuller, *Children and the Movies: Media Influence and the Payne Fund Controversy* (New York: Cambridge University Press, 1996); Ellen Wartella and Michael Robb, "Historical and Recurring Concerns about Children's Use of the Mass Media," in *The Handbook of Children, Media, and Development*, ed. Sandra L. Calvert and Barbara J. Wilson (Malden, MA: Wiley-Blackwell, 2011), 7–27.

39. Frances Norene Ahl, "The Use of Audio-Visual Aids in an International Relations Class," *Educational Screen* 24, no. 3 (1945): 102–104.

40. Alison Griffiths, *Wondrous Difference: Cinema, Anthropology and Turn-of-the-Century Visual Culture* (New York: Columbia University Press, 2002); Robert W. Rydell, *World of Fairs: The Century-of-Progress Expositions* (Chicago: University of Chicago Press, 1993).

41. Saettler, *Evolution of American Educational Technology*, 128–136, 166.

42. Jean Ramsey, "Visual Education on Wheels," *Visual Education* 3, no. 6 (1922): 285–287, 305; Saettler, *Evolution of American Educational Technology*, 128.

43. Ramsey, "Visual Education on Wheels," 287.

44. Carl G. Rathmann, "Educational Museum of the St. Louis Public Schools," *US Bureau of Education Bulletin* 48, no. 622 (1914): 7, 11. On school museum collections containing "a little bit of the reality of the world about which the pupil is reading

and studying," see J. Paul Goode, "Scope and Outlook of Visual Education," *Visual Education* 1, no. 2 (1920): 7.

45. Carl G. Rathmann, "Visual Education and the St. Louis School Museum," *US Bureau of Education Bulletin* 39 (1924): 12.

46. Griffiths, *Wondrous Difference*, 14–15.

47. Rathmann, "Visual Education and the St. Louis School Museum," 2, 7–8.

48. James Clifford, "Objects and Selves: An Afterword," in *Objects and Others: Essays on Museums and Material Culture*, ed. George W. Stocking Jr. (Madison: University of Wisconsin Press, 1985), 236–246; Susan Stewart, *On Longing: Narratives of the Miniature, the Gigantic, the Souvenir, the Collection* (Durham, NC: Duke University Press, 1984); Tony Bennett, *The Birth of the Museum: History, Theory, Politics* (New York: Routledge, 1995).

49. Rathmann, "Educational Museum of the St. Louis Public Schools," 16.

50. Amelia Milone, "Realia as Applied in the Italian Classroom," *Modern Language Journal* 22, no. 5 (1938): 353.

51. J. Alan Pfeffer, "Realia in Modern Language Instruction," *German Quarterly* 10, no. 1 (1937): 1.

52. Lillian L. Stroebe, "The Real Knowledge of a Foreign Country," *Modern Language Journal* 4, no. 6 (1920): 294.

53. Findlay and Perricone, *WPA Museum Extension Project.*

54. Marian Evans, "Visualization of Today's High-School Curriculum," *Junior-Senior High School Clearing House* 9, no. 6 (1935): 375-376.

55. Maurice Hunt, "Visual and Other Aids," *Social Studies* 35, no. 3 (1944): 128–130; Ella Huntting put it this way: "Old texts may be cut and pasted together to answer new problems." Ella Huntting, "The School Neighborhood," *Journal of Geography* 26 (1927): 237.

56. Fannie Fern Andrews, "Work of the American School Peace League for May," *Advocate of Peace*, 72, no. 6 (1910): 149; Charles E. Luminati, "21 Devices for Teaching Current Events," *Clearing House* 15, no. 1 (1940): 37; Frederick K. Branom, *The Teaching of the Social Studies in a Changing World* (New York: W. H. Sadlier, Inc., 1942), 99; R. S. Ihlenfeldt, "Bulletin Boards and Pupil Learning," *See and Hear* 1, no. 8 (1946): 48–53; National Education Association, *Education for International Understanding in American Schools: Suggestions and Recommendations* (Washington, DC: National Education Association of the United States, 1948), 119. The idea of using bulletin boards to interact with maps and identify hot spots around the world may have arisen from the unit on identifying "sore spots" (conflict zones or territories disputed among nations) developed by Harold Rugg, Earle Rugg, and Emma

Schewppe in their influential Social Studies Pamphlets series. See "Map Studies of the Storm Centers of the World," in *How Nations Live Together*, Social Studies Pamphlets, 1923, 1, 27–59.

57. James N. Emery, "Sources of Visual Aids at Moderate Cost," *Educational Screen* 3, no. 6 (1924): 212–213.

58. Joseph Burton Vasché, "10 Ideas for Timely Teaching of 1943 Social Studies: Stanislaus County Schools' Plan of Action," *The Clearing House* 17, no. 8 (1943): 472; Luminati, "21 Devices for Teaching Current Events," 37.

59. Albert A. Orth, "The Bulletin Board on Special Days," *Social Studies* 29, no. 4 (1938): 172–173.

60. Etienne, "Opaque Projector."

61. Ahl, "Use of Audio-Visual Aids," 103–104.

62. Spencer Lens Company, "Knowledge, up to the Minute," *Educational Screen* 24, no. 5 (1945): 199.

63. Ellen Gruber Garvey, "Scizzoring and Scrapbooks: Nineteenth-Century Reading, Remaking, and Recirculating," in *New Media, 1740–1915*, ed. Lisa Gitelman and Geoffrey Pingree (Cambridge, MA: MIT Press, 2004), 207–208; Katie Day Good, "From Scrapbook to Facebook: A History of Personal Media Assemblage and Archives," *New Media and Society* 15, no. 4 (2013): 557–573; Georgie B. Collins, "The Scrapbook as a Visual Aid in Health Education," *Visual Education*, 1922, 132–137, 172; Kittel, "Pictures and Scrap Books"; J. P. Givler, "Picture Collections: How to Rescue, Organize, and Store Them," *ALA Bulletin* 33, no. 1 (January 1, 1939): 29–50; John Paul Givler, "An Undeveloped Mine of Materials for Visual Education," *Educational Screen* 19, no. 2 (1940): 53–55.

64. Maurice Hunt, "Visual and Other Aids," *Social Studies* 35, no. 3 (1944): 128–130. See also George E. Mark, "Use of the Newspaper," *Junior-Senior High School Clearing House* 8, no. 5 (1934): 310–311; Ruth Messenger, "Assimilative Material," *English Journal* 23, no. 1 (1934): 58–63.

65. Mark, "Use of the Newspaper," 310–311.

66. Joel H. Spring, *Images of American Life: A History of Ideological Management in Schools, Movies, Radio, and Television* (Albany: SUNY Press, 1992), 123–126.

67. Edgar Dale, "Radio-TV Institute," June 14, 1963, 1, Edgar Dale Papers, box 3, folder 3/19, Ohio State University; Edgar Dale, "Let Us Raise the Standard," *News Letter* 10, no. 2 (1944).

68. Hood-Hanchey, "Transformation of Experience," 11–21.

69. Elmer A. Winters, "Man and His Changing Society: The Textbooks of Harold Rugg," *History of Education Quarterly* 7, no. 4 (December 1, 1967): 497; Harold

Rugg, Earle Rugg, and Emma Schewppe, *How Nations Live Together*, Social Studies Pamphlets, 1923, 12; LaGarrett J. King, Christopher Davis, and Anthony L. Brown, "African American History, Race, and Textbooks: An Examination of the Works of Harold O. Rugg and Carter G. Woodson," *Journal of Social Studies Research* 36, no. 4 (2012): 359–386.

70. Karen L. Riley and Barbara Slater Stern, "'A Bootlegged Curriculum': The American Legion versus Harold Rugg," *International Journal of Social Education* 18, no. 2 (2004): 62–72; Patricia Albjerg Graham, *Schooling America: How the Public Schools Meet the Nation's Changing Needs* (New York: Oxford University Press, 2007), 67–70.

71. Frank N. Freeman, "A Scientific Study of Visual Education," *Journal of Educational Research* 10, no. 5 (1924): 375–385.

72. Edgar Dale, "Associations with W. W. Charters," 1953, Edgar Dale Papers, box 7, folder 2, Ohio State University; Saettler, *Evolution of American Educational Technology*, 145–147.

73. Benjamin Harrison Darrow, *Radio Trailblazing: A Brief History of the Ohio School of the Air and Its Implications for Educational Broadcasting* (Columbus, OH: College Book Company, 1940), 18–19. For a history of the Payne Fund's involvement in educational broadcasting, see Robert McChesney, "The Payne Fund and Radio Broadcasting," in *Children and the Movies: Media Influence and the Payne Fund Controversy*, ed. Garth S. Jowett, Ian C. Jarvie, and Kathryn H. Fuller (New York: Cambridge University Press, 1996), 303–335.

74. On sales of *How to Appreciation Motion Pictures*, see Acland, "American AV," 397. Between 1934 and 1936, the book was used in a five-state study (North Carolina, Ohio, Iowa, California, and Connecticut) led by the National Committee on Motion Picture Appreciation, chaired by US commissioner of education George Zook and supported by the Payne Fund. Dale was also invited by the state of Pennsylvania to lead a study in motion picture appreciation in forty-five cities. See Edgar Dale, "Teaching Motion Picture Appreciation: An Account of a Series of Demonstrations in Forty-Five Selected Pennsylvania Cities," Bureau of Educational Research, 1936, Edgar Dale Papers, box 7, folder 137, Ohio State University.

75. Wartella and Robb, "Historical and Recurring Concerns about Children's Use of the Mass Media."

76. McChesney, "Payne Fund and Radio Broadcasting"; Saettler, *Evolution of American Educational Technology*, 209–212.

77. Eugene E. Leach, "Tuning Out Education: The Cooperation Doctrine in Radio, 1922–38," *Current*, August 1983, 2.

78. The decade saw increasing collaboration between educators and the film and radio industries, as the industries grew in power and educators' hopes faded that

they would be legislatively reformed. McChesney, "Payne Fund and Radio Broadcasting," 314; Josh Shepperd, "Infrastructure in the Air: The Office of Education and the Development of Public Broadcasting in the United States, 1934–1944," *Critical Studies in Media Communication* 31, no. 3 (May 27, 2014): 230–243.

79. John Nichols, "Countering Censorship: Edgar Dale and the Film Appreciation Movement," *Cinema Journal* 46, no. 1 (2006): 3–22.

80. Edgar Dale, "The Motion Picture and Intergroup Relationships," *Public Opinion Quarterly* 2, no. 1 (1938): 40.

81. Jowett, Jarvie, and Fuller, *Children and the Movies*, 113.

82. Edgar Dale, *The Content of Motion Pictures* (New York: Macmillan, 1935), 41, 60–61, 54, 63. See also Edgar Dale, "The Movies and Race Relations," *Crisis* (1937): 294, 296.

83. Dale, "Movies and Race Relations," 315.

84. Edgar Dale, "Motion Pictures and the War," *Educational Screen* 21, no. 6 (1942): 213–214.

85. Dale, "Movies and Race Relations," 294; Dale, "Motion Picture and Intergroup Relationships," 42.

86. Alice Sterner of Columbia University's Teachers College was another leader in developing educational materials for radio appreciation. She authored several studies and a pamphlet. Alice P. Sterner, *Skill in Listening*, ed. Lennox Grey, NCTE Pamphlets on Communication (Chicago: National Council of Teachers of English, 1944). See also Max J. Herzberg, "Tentative Units in Radio Program Appreciation," *English Journal* 24, no. 7 (1935): 545–555; Samuel G. Gilburt, "Radio Appreciation: A Plea and a Program," *English Journal* 32, no. 8 (1943): 431–435; David Goodman, *Radio's Civic Ambition: American Broadcasting and Democracy in the 1930s* (New York: Oxford University Press, 2011).

87. Dale, *Audio-Visual Methods in Teaching*, 259–265, 449.

88. I. Keith Tyler, "How to Judge a Radio Program," *Scholastic* 27, no. 14 (January 11, 1936); I. Keith Tyler, *High School Students Talk It Over* (Columbus: Bureau of Educational Research, Ohio State University, 1937).

89. Edgar Dale, "Unlicensed Teachers: Radio, Movies and Press," *Education Digest* 6, no. 7 (1941): 21–23.

90. Edgar Dale, *How to Read a Newspaper* (Chicago: Scott, Foresman and Company, 1941); Stella S. Center, "The Responsibility of Teachers of English in Contemporary American Life," *English Journal* 22, no. 2 (1933): 102–103.

91. Dale, *How to Read a Newspaper*, 19, 123–124.

92. Institute for Propaganda Analysis, "Press Release," January 15, 1941, Institute for Propaganda Analysis Papers, box 1, file 5, New York Public Library; J. Michael Sproule, *Propaganda and Democracy: The American Experience of Media and Mass Persuasion* (Cambridge: Cambridge University Press, 1997), 129–177.

93. Howard Cummings, "Teaching Propaganda Analysis: Clayton High School Methods Rank High among Those of 400 Schools," *Clearing House* 13, no. 7 (1939): 394–398; Institute for Propaganda Analysis, *Propaganda: How to Recognize It and Deal with It* (New York: Institute for Propaganda Analysis, Inc., 1938), 3; Edgar Dale and Norma Wynne Vernon, *Propaganda Analysis: An Annotated Bibliography* (Columbus: Bureau of Educational Research, Ohio State University, 1940).

94. Sproule, *Propaganda and Democracy*, 137.

95. Renee Hobbs and Sandra McGee, "Teaching about Propaganda: An Examination of the Historical Roots of Media Literacy," *Journal of Media Literacy Education* 6, no. 2 (2014): 56–67.

96. Clyde R. Miller, "Propaganda Analysis," *Publications of the Institute for Propaganda Analysis, Inc.* 1 (1938): iv; Institute for Propaganda Analysis, *Propaganda*, 3; Hadley Cantril, "Propaganda Analysis," *English Journal* 27, no. 3 (March 1, 1938): 217.

97. Miller, "Propaganda Analysis," iii; Cantril, "Propaganda Analysis," 217; Institute for Propaganda Analysis, *Propaganda*, 16.

98. Miller, "Propaganda Analysis," iv; Anonymous, "Monthly Propaganda Analysis for Schools," *Clearing House* 12, no. 2 (October 1, 1937): 113.

99. Clyde R. Miller, "Propaganda and Press Freedom," *English Journal* 28, no. 10 (1939): 824.

100. Institute for Propaganda Analysis, "Analysis of Propaganda Suspended for Duration," *New York Herald Tribune*, October 31, 1941, Institute for Propaganda Analysis Papers, box 1, file 5, New York Public Library. Miller protested the closure of the IPA and insisted that it was a lack of funding, not the needs of wartime, that caused it to close. He maintained that during the war, the "clear thinking" promoted by the IPA was sorely needed. Clyde R. Miller, "Letter to Clyde Beals," November 19, 1941, Institute for Propaganda Analysis Papers, box 1, folder: "Miller, Clyde R.," New York Public Library; Clyde R. Miller, "Letter to Alfred McClung Lee," August 9, 1950, Institute for Propaganda Analysis Papers, box 1, folder: "Miller, Clyde R.," New York Public Library; Sproule, *Propaganda and Democracy*, 170.

101. Edward Fiess, "Language and Morals," *College English* 6, no. 5 (February 1, 1945): 271; Cummings, "Teaching Propaganda Analysis"; Katherine Sommers, "Propaganda, an English Project," *English Journal* 27, no. 7 (1938): 598–600; Hobbs and McGee, "Teaching about Propaganda."

102. Clyde R. Miller, "Letter to Alfred McClung Lee," August 26, 1947, Institute for Propaganda Analysis Papers, box 1, folder: "I. P. A.," New York Public Library.

103. Edgar Dale, "Motion Pictures and the War," *Educational Screen* 21, no. 6 (1942): 213–217.

104. Bess Goodykoontz, "Propaganda: What It Is; How It Works; What to Do about It," *Pi Lambda Theta Journal* 17, no. 1 (1938): 6–9; Henry H. Hill, *Intercultural Education through Administration and Guidance* (Pittsburgh, PA: Board of Public Education, 1944); Joseph Gallant, "An Intercultural Curriculum," *English Journal* 33, no. 7 (1944): 382; Nathan Miller, "Listening to Learn," *See and Hear* (1945): 71–75; Leon S. Kaiser and Leanora S. Ratner, "A Project in Inter-Racial and Inter-Faith Education: Finding the Road to Peace and Victory," *Elementary English Review* 21, no. 3 (1944): 81–88. On the association between propaganda analysis and citizenship education, see Stanley E. Dimond, "Catch-all Citizenship," *Clearing House* 25, no. 5 (1951): 270.

105. Dale and Vernon, *Propaganda Analysis*, i; Miller, "Propaganda and Press Freedom."

106. Edgar Dale, "Sponge-Minded or Critical?," *Social Progress* 30, no. 5 (December 1939): 11–12.

Chapter 4

1. Ella Lyman Cabot, Fannie Fern Andrews, Fanny E. Coe, Mabel Hill, and Mary McSkimmon, *A Course in Citizenship* (Boston: Houghton Mifflin, 1914), 283–284; Adeline Mair, "Quelling a 'Class of Babel' by a Unit on Foreign Cultures," *Clearing House* 13 (1939): 265–266.

2. Megan E. Geigner, "Performing the Polish-American Patriot: Civic Performance and Hyphenated Identity in World War I Chicago," *Theatre History Studies* 34, no. 1 (2015): 59–78.

3. Bruno Lasker, *Race Attitudes in Children* (New York: Henry Holt and Company, 1929), 223.

4. Michael C. Johanek and John L. Puckett, *Leonard Covello and the Making of Benjamin Franklin High School: Education as If Citizenship Mattered* (Philadelphia: Temple University Press, 2007), 27; Lyda Judson Hanifan, *The Community Center* (Boston: Silver, Burdett and Company, 1920), vii; Minnesota Public Library Commission, "Social Centers," in *Library Notes and News* (Roseville: Minnesota Department of Education, 1915), 4:51.

5. Margaret Woodrow Wilson, "Getting Together: What Every Community Can Do to Help in the War," *Ladies' Home Journal*, December 1917; Edward W. Stevens, "Social Centers, Politics, and Social Efficiency in the Progressive Era," *History of*

Education Quarterly 12, no. 1 (1972): 16–33; Dudley Grant Hayes, "Public School Community Center Work with Reel and Slide," *Moving Picture Age* 3, no. 7 (1920): 11.

6. John Dewey, "The School as Social Center," *Elementary School Teacher* 3, no. 2 (1902): 73–86; Johanek and Puckett, *Leonard Covello*, 24–26.

7. Julie A. Reuben, "Beyond Politics: Community Civics and the Redefinition of Citizenship in the Progressive Era," *History of Education Quarterly* 37, no. 4 (1997): 399–420.

8. Edwin Wesley Adams, *A Community Civics: A Text-Book in Loyal Citizenship* (New York: Charles Scribner's Sons, 1920); J. Lynn Barnard, F. W. Carrier, Arthur William Dunn, and Clarence D. Kingsley, "The Teaching of Community Civics," *US Bureau of Education Bulletin*, no. 23 (1915).

9. Arthur William Dunn and Hannah Margaret Harris, *Citizenship in School and Out* (Boston: D. C. Heath and Company, 1919), 2; Arthur William Dunn, *Community Civics and Rural Life* (D. C. Heath and Company, 1920), 89, 96; Arthur William Dunn, *Community Civics for City Schools* (D. C. Heath and Company, 1921); Reuben, "Beyond Politics." After the war, Dunn continued to advocate for internationalism in education as the national director of the American Junior Red Cross. Julia F. Irwin, *Making the World Safe: The American Red Cross and a Nation's Humanitarian Awakening* (New York: Oxford University Press, 2013), 269–270. The riskiness of advocating for teaching internationalism during the war, and need to continuously couch it in appeals to national loyalty, is evident in the anonymously authored "Professor X.," "Some Thoughts on Nationalism and Internationalism," *History Teacher's Magazine* 9, no. 6 (1918): 334.

10. Geigner, "Performing the Polish-American Patriot"; Charlotte Canning, *The Most American Thing in America: Circuit Chautauqua as Performance* (Iowa City: University of Iowa Press, 2005).

11. David Glassberg, *American Historical Pageantry: The Uses of Tradition in the Early Twentieth Century* (Chapel Hill: University of North Carolina Press Books, 1990), 64; Canning, *Most American Thing in America*, 34.

12. Gertrude Jones, *Commencement* (New York: A. S. Barnes and Company, 1929), 38–43; Glassberg, *American Historical Pageantry*, 63.

13. George McReynolds, "The Centennial Pageant for Indiana: Suggestions for Its Performance," *Indiana Magazine of History* 11, no. 3 (1915): 248–271.

14. Ralph Davol, *A Handbook of American Pageantry* (Taunton, MA: Davol Publishing Co., 1914), 77; John Higham, *Hanging Together: Unity and Diversity in American Culture* (New Haven, CT: Yale University Press, 2001), 223.

15. Davol, *Handbook of American Pageantry*, 84–85.

16. Lotta A. Clark, "Pageantry in America," *English Journal* 3, no. 3 (1914): 151.

17. Henry Lester Smith and Sherman Gideon Crayton, "Tentative Program for Teaching World-Friendship and Understanding in Teacher Training Institutions and in Public Schools for Children Who Range from Six to Fourteen Years of Age," *Bulletin of the Indiana University School of Education* 5, no. 5 (1929): 47.

18. James N. Emery, "Sources of Visual Aids at Moderate Cost," *Educational Screen* 3, no. 6 (1924): 211–213; George C. Kyte, "Enriching Learning through the Use of Visual Aids," *Educational Screen* (1923).

19. John Dewey and Evelyn Dewey, *Schools of To-Morrow* (New York: E. P. Dutton and Company, 1915).

20. Amalie Hofer, "The Significance of Recent National Festivals in Chicago," in *Proceedings of the Second Annual Playground Congress* (New York: Playground Association of America, 1908), 2:74–86; Victor Greene, "Dealing with Diversity: Milwaukee's Multiethnic Festivals and Urban Identity, 1840–1940," *Journal of Urban History* 31, no. 6 (2005): 820–849; Jeffrey Mirel, *Patriotic Pluralism: Americanization Education and European Immigrants* (Cambridge, MA: Harvard University Press, 2010), 210.

21. Joel H. Spring, *Images of American Life: A History of Ideological Management in Schools, Movies, Radio, and Television* (Albany: SUNY Press, 1992), 31–32; Karen L. Riley and Barbara Slater Stern, "'A Bootlegged Curriculum': The American Legion versus Harold Rugg," *International Journal of Social Education* 18, no. 2 (2004): 62–72; John Higham, *Strangers in the Land: Patterns of American Nativism, 1860–1965* (New Brunswick, NJ: Rutgers University Press, 2011), 236–237.

22. F. E. King, *The Pageant of Escanaba and Correlated Local History* (Grand Rapids, MI, 1917), 371.

23. King, *Pageant of Escanaba*, 342–343.

24. Mirel, *Patriotic Pluralism*, 48; Adams, *Community Civics*; Hofer, "Significance of Recent National Festivals"; Mair, "Quelling a 'Class of Babel.'"

25. "Melting Pot Ceremony at Ford English School, July 4, 1917," Henry Ford, https://www.thehenryford.org/collections-and-research/digital-collections/artifact/254569/#slide=gs-213137. The Ford "Melting pot" ceremony was described in detail by Walter Lippmann, who regarded Americanization programs as largely superficial exercises that resulted in the "substitution of American for European stereotypes." In *Public Opinion* (New York: Harcourt, Brace and Co., 1922), chapter 6.

26. Francis J. Brown, "Sociology and Intercultural Understanding," *Journal of Educational Sociology* 12, no. 6 (1939): 328–329.

27. Lasker, *Race Attitudes in Children*, 295.

28. Diana Selig, *Americans All: The Cultural Gifts Movement* (Cambridge, MA: Harvard University Press, 2008), 295; Matthew Frye Jacobson, *Whiteness of a Different Color: European Immigrants and the Alchemy of Race* (Cambridge, MA: Harvard University Press, 1998).

29. Susan Zeiger, "Teaching Peace: Lessons from a Peace Studies Curriculum of the Progressive Era," *Peace and Change* 25, no. 1 (2000): 52–69.

30. American School Peace League, "First Annual Report of the American School Peace League," October 1909, 1, American School Citizenship League Collected Records, box 1, folder 1, Swarthmore Peace Collection.

31. For more information on how Peace Day came to be celebrated in schools, with endorsements from the US Bureau of Education as early as 1906 and resources from the agency in 1912–1914, see Fannie Fern Andrews, "The Teacher an Agent of International Goodwill," in *Proceedings of the Sixty-Fifth Annual Meeting of the National Education Association*, July 3, 1927, 428–429. See also Charles F. Howlett and Ian M. Harris, *Books, Not Bombs: Teaching Peace since the Dawn of the Republic* (Charlotte, NC: Information Age Publishing, 2010), 63–70; Zeiger, "Teaching Peace."

32. Andrews, "Teacher an Agent of International Goodwill," 429; Fannie Fern Andrews, "Third Annual Report of the American School Peace League," 1911, 44, American School Citizenship League Collected Records, box 1, folder 2, Swarthmore Peace Collection.

33. Cabot, Andrews, Coe, Hill, and McSkimmon, *Course in Citizenship*, xix, 84; American School Citizenship League, "Aims," 1921, 3, American School Citizenship League Collected Records, box 3, folder 2, Swarthmore Peace Collection.

34. In 1912, the ASPL was said to have circulated a hundred thousand bulletins, containing suggestions for programs and materials on "The Promotion of Peace," to schools at the request of US commissioner of education P. P. Claxton. Fannie Fern Andrews, "An Eleven-Year Survey of the Activities of the American School Citizenship League from 1908 to 1919," 1919, 15, American School Citizenship League Collected Records, box 2, folder "1918–19," Swarthmore Peace Collection; American School Peace League, "First Annual Report of the American School Peace League," 13–14; Lyra Trueblood, "How the Eighteenth of May Came to Be Observed as Peace Day," 1915, American School Citizenship League Collected Records, box 1, folder "1915," Swarthmore Peace Collection.

35. Beulah Marie Dix, *A Pageant of Peace*, 1915, foreword, American School Citizenship League Collected Records, box 1, folder "1915," Swarthmore Peace Collection.

36. Dix, *Pageant of Peace*, 5–6.

37. Anonymous, "Pilgrim Festival," *Visual Education* 1, no. 2 (1920): 37; Anonymous, "Armistice Day Program," *Instructor* 29 (1919): 50; Jones, *Commencement*;

Elizabeth Miller Lobingier and John Leslie Lobingier, *Educating for Peace* (Boston: Pilgrim Press, 1930), 177.

38. Dix, *Pageant of Peace*, 17.

39. Andrews, "Eleven-Year Survey," 20.

40. Zeiger, "Teaching Peace," 53; Howlett and Harris, *Books, Not Bombs*.

41. Zeiger, "Teaching Peace," 63, 53.

42. Andrews, "Teacher an Agent of International Goodwill," 428–430.

43. E. Estelle Downing, "International Good Will through the Teaching of English," *English Journal* 14, no. 9 (1925): 677.

44. Ruth A. Barnes, "Developing International-Mindedness in Junior High School," *English Journal* 22, no. 6 (1933): 476–481; Ida T. Jacobs, "The Classroom a Laboratory in International Relations," *English Journal* 27, no. 8 (1938): 666–673; Mair, "Quelling a 'Class of Babel'"; Henrietta F. Hafemann, "The Senn High School Intercultural Relations Laboratory," *See and Hear* (1946): 36–37.

45. Evaline Dowling, ed., *World Friendship: A Series of Articles Written by Some Teachers in the Los Angeles Schools and by a Few Others Who Are Likewise Interested in the Education of Youth* (Los Angeles: Committee on World Friendship, 1928), 78; Suzanne Croizat Borghei, "Internationalism at the Grassroots: Los Angeles and Its City Schools, 1916–1953" (PhD diss., University of Southern California, 1995).

46. Dowling, *World Friendship*, 78.

47. Jay Walz, "Pupils Drink Punch; Stage Peace Play," *Washington Post*, April 26, 1936.

48. Zoe Burkholder, *Color in the Classroom: How American Schools Taught Race, 1900–1954* (New York: Oxford University Press, 2011); Selig, *Americans All*; Susan Schulten, *The Geographical Imagination in America, 1880–1950* (Chicago: University of Chicago Press, 2001).

49. William George Carr, *Education for World-Citizenship* (Palo Alto, CA: Stanford University Press, 1928); Bertha M. Bartholomew and C. L. Kulp, "Public-School Activities Designed to Develop Wholesome Nationalism and International Understanding," *Journal of Educational Sociology* 9, no. 7 (1936): 408–410; Henry Lester Smith and Peyton Henry Canary, "Some Practical Efforts to Teach Good Will," *Bulletin of the Indiana University School of Education* 6, no. 4 (1935): 3–169; I. L. Kandel, "International Understanding and the Schools," *NASSP Bulletin* 9, no. 1 (1925): 36–41; I. L. Kandel and Guy Montrose Whipple, eds., *International Understanding through the Public-School Curriculum* (Bloomington, IL: Public School Publishing Company, 1937); Augustus O. Thomas, "Education as a World Problem," *Journal of the National Education Association* 11, no. 8 (1922): 314–316; Anonymous, "International

Goodwill through Education," *Journal of the National Education Association* 11, no. 2 (1922): 45–46.

50. National Education Association, *World Conference on Education* (Oakland, CA: National Education Association, 1923), 19–20; Anonymous, "World Goodwill Day to Be Widely Observed," *Federal Council Bulletin*, 1927, 8; James F. Abel, "Goodwill Day—May 18," *School Life* 18, no. 9 (1933): 169.

51. Dowling, *World Friendship*; Spencer Stoker, *The Schools and International Understanding* (Chapel Hill: University of North Carolina Press, 1933); Ruth A. Barnes, "Developing International-Mindedness in Junior High School," *English Journal* 22, no. 6 (1933): 476–481; Smith and Canary, "Some Practical Efforts to Teach Good Will"; Edgar G. Johnston, "A Contribution to International Attitudes," *Journal of Educational Sociology* 9, no. 7 (1936): 421–425; Bartholomew and Kulp, "Public-School Activities"; Kenneth E. Gell, "What the Rochester Schools Are Doing about Internationalism," *Journal of Educational Sociology* 9, no. 7 (March 1, 1936): 397–407; Malcolm B. Keck, "National and International Understanding," *Journal of Educational Sociology* 9, no. 7 (1936): 418–420.

52. Stephen Stanton Myrick, "A Step toward Internationalism," *Los Angeles School Journal*, 1919; Dowling, *World Friendship*, 186; Albert John Murphy, *Education for World-Mindedness* (New York: Abingdon Press, 1931), 299–301.

53. Henry G. Wellman, "School and Community for International Understanding," *Journal of Educational Sociology* 9, no. 7 (1936): 411–417; William M. Keith, *Democracy as Discussion: Civic Education and the American Forum Movement* (Lanham, MD: Lexington Books, 2007).

54. William Dow Boutwell, "Education in Good-Will: Schools in Every State Are Seeking to Promote International Peace," *New York Times*, April 19, 1936.

55. Fred Turner, *The Democratic Surround: Multimedia and American Liberalism from World War II to the Psychedelic Sixties* (Chicago: University of Chicago Press, 2013), 17–20.

56. Robert Frederick, "An Investigation into Some Social Attitudes of High School Pupils," *School and Society* 25, no. 640 (April 2, 1927): 410–412. DuBois, echoing the writer Louis Adamic, remarked that "there are 30 million new Americans of the second generation with such inferiority complexes regarding their own cultural backgrounds that they cannot function as the kind of citizens that a democracy needs." Rachel Davis DuBois, "Our Enemy—the Stereotype," *Progressive Education* 12, no. March (1935): 147.

57. Frederic M. Thrasher, "Social Attitudes of Superior Boys in an Interstitial Community," in *Social Attitudes*, ed. Kimball Young (New York: Henry Holt, 1931), 237.

58. Lasker, *Race Attitudes in Children*, 115–135, 298–299; Selig, *Americans All*, 14–15.

59. Ruth C. Peterson and L. L. Thurstone, *Motion Pictures and the Social Attitudes of Children* (New York: Macmillan, 1933); Werrett Wallace Charters, *Motion Pictures and Youth: A Summary* (New York: Macmillan, 1933).

60. Murphy, *Education for World-Mindedness*, 46–47.

61. Helen K. Champlin, "Will Our Children Outlaw War?," *Parents' Magazine* (July 1933): 14–15, 50; Bruno Lasker, "How Children Acquire Race Prejudices," *Parents' Magazine* 3, no. 3 (March 1928): 23–24, 42; Rachel Dunaway Cox, "World Friendship among Children," *Parents' Magazine* (July 1930): 16–18, 52. For a detailed history of parent education in teaching tolerance, see Selig, *Americans All*, 39–67.

62. Elizabeth B. Wisdom, "International Friendship in Children's Reading," *Elementary English Review* 2, no. 5 (1925): 158.

63. Lasker, *Race Attitudes in Children*, 223–225.

64. Lobingier and Lobingier, *Educating for Peace*, 152; Cox, "World Friendship among Children," 17; Patricia Faith Appelbaum, *Kingdom to Commune: Protestant Pacifist Culture between World War I and the Vietnam Era* (Chapel Hill: University of North Carolina Press, 2009), 172.

65. Ella Huntting, "The School Neighborhood," *Journal of Geography* 26 (1927): 235–236; Ella Adeline Busch, "How We Learn to Understand Our Neighbors," *Hispania* 6 (1923): 205–213.

66. Joseph S. Roucek, "Future Steps in Cultural Pluralism," *Journal of Educational Sociology* 12, no. 8 (1939): 502–503; Mair, "Quelling a 'Class of Babel.'"

67. Murphy, *Education for World-Mindedness*, 263, 267–268.

68. Murphy, for example, suggested a skit in which American and Japanese children exchanged tea and manufactured goods so that "the delightful characteristics of the Japanese could be acted out." Murphy, *Education for World-Mindedness*, 264–265. See a similar reference to intercultural skits in Gell, "What the Rochester Schools Are Doing."

69. Anna Pettit Broomell, ed., "A Boot Is a League of Nations," in *The Children's Story Caravan* (Philadelphia: J. B. Lippincott and Company, 1935), 220–224. Another version, "A Shoe, a League of Nations," is described in Appelbaum, *Kingdom to Commune*, 173–174.

70. The pageant was recommended for public school use in E. Estelle Downing, "For International Good-Will Day," *English Journal* 16, no. 3 (1927): 230; Carr, *Education for World-Citizenship*, 191–193. On the history of the Ku Klux Klan slogan, see Ralph McGill, *The South and the Southerner*, 3rd ed. (Athens: University of Georgia Press, 1992), 133; "America for Americans" (Ku Klux Klan, n.d.), Kansas Memory, Kansas Historical Society, http://www.kansasmemory.org/item/214406.

71. Murphy, *Education for World-Mindedness*, 261.

72. Florence Brewer Boeckel, *Through the Gateway* (New York: Macmillan, 1928), 67–73.

73. Selig, *Americans All*, 69; Shafali Lal, "1930s Multiculturalism: Rachel Davis DuBois and the Bureau for Intercultural Education," *Radical Teacher*, no. 69 (2004): 18–22; Benjamin Looker, "Microcosms of Democracy: Imagining the City Neighborhood in World War II-Era America," *Journal of Social History* 44, no. 2 (2010): 351–378; Thomas Fallace, *Race and the Origins of Progressive Education, 1880–1929* (New York: Teachers College Press, 2015), 139.

74. Selig, *Americans All*, 73.

75. DuBois, "Our Enemy," 147; Rachel Davis DuBois, "Sharing Culture Values," *Journal of Educational Sociology* 12, no. 8 (1939): 485.

76. Selig, *Americans All*, 74.

77. Rachel Davis DuBois, "Practical Problems of International and Interracial Education," *Junior-Senior High School Clearing House* 10, no. 8 (1936): 486–487; Roucek, "Future Steps in Cultural Pluralism," 500, 502. As Dewey remarked in a letter to Horace Kallen, who coined the term "cultural pluralism" in 1924, embracing differences was the essence of American democracy. "I never did care for the melting pot metaphor, but genuine assimilation *to one another*—not to Anglosaxondom—seems to be essential to America." Quoted in Selig, *Americans All*, 94.

78. Lal, "1930s Multiculturalism," 21; Selig, *Americans All*, 6.

79. Selig, *Americans All*, 77.

80. DuBois, "Our Enemy," 146.

81. Rachel Davis DuBois, "A Program for Education in Worldmindedness: A Series of Assembly Programs Given by Students of Woodbury High School," Women's International League for Peace and Freedom, 1928, Rachel Davis DuBois Papers, 1920–1933, box 4, folder 168, Friends Historical Library of Swarthmore College; Rachel Davis DuBois, "Pioneers of the New World: Series III of the Assembly Programs Given by Students of Woodbury, New Jersey High School," Women's International League for Peace and Freedom, 1930, Rachel Davis DuBois Papers, 1920–1933, box 4, folder 168, Friends Historical Library of Swarthmore College; Rachel Davis DuBois, "Dawn Edge, Woodstown, N. J.," August 2, 1989, Rachel Davis DuBois Papers, 1920–1933, box 8, folder 149, Friends Historical Library of Swarthmore College.

82. She cites *A Course in Citizenship and Patriotism* as a "very good" resource. DuBois, "A Program for Education in Worldmindedness," 35.

83. Selig, *Americans All*, 76.

84. Rachel Davis DuBois, "Building Tolerant Attitudes in High-School Students," *Crisis* 38 (1931): 334.

85. Rachel Davis DuBois, "Developing Sympathetic Attitudes toward Peoples," *Journal of Educational Sociology* 9, no. 7 (1936): 392.

86. Selig, *Americans All*, 90–92.

87. DuBois, "Developing Sympathetic Attitudes toward Peoples," 390, quoted in Selig, *Americans All*, 77; Rachel Davis DuBois, "Shall We Emotionalize Our Students?," *Friends Intelligencer* 3 (1932): 973–974.

88. Rachel Davis DuBois, "Pioneers of the New World"; DuBois, "Building Tolerant Attitudes in High-School Students," 334.

89. DuBois, "Developing Sympathetic Attitudes toward Peoples," 392; DuBois, "Building Tolerant Attitudes in High-School Students."

90. DuBois, "Our Enemy," 147.

91. DuBois, "Our Enemy," 148; Rachel Davis DuBois, "The Role of Home Economics in Intercultural Education," *Journal of Home Economics* 30, no. 3 (1938): 147.

92. Lal, "1930s Multiculturalism," 19; Selig, *Americans All*, 99–103.

93. Service Bureau for Education in Human Relations, "Changing Race Attitudes: A Course for Teachers, Social Workers and Community Leaders," 1934, Rachel Davis DuBois Papers, 1920–1933, box 4, folder 168, Friends Historical Library of Swarthmore College.

94. DuBois, "Dawn Edge, Woodstown, N. J."

95. Mirel, *Patriotic Pluralism*, 162.

96. US Department of the Interior, *Public Affairs Pamphlets*, vol. 3 (Washington, DC: US Government Printing Office, 1937); Keith, *Democracy as Discussion*, 295–297; Selig, *Americans All*, 246.

97. DuBois, "Our Enemy," 149–150.

98. Barbara Dianne Savage, *Broadcasting Freedom: Radio, War, and the Politics of Race, 1938–1948* (Chapel Hill: University of North Carolina Press, 1999), 24; Selig, *Americans All*, 243.

99. Institute for Propaganda Analysis, "Father Coughlin: Priest and Politician," *Propaganda Analysis: A Bulletin to Help the Intelligent Citizen Detect and Analyze Propaganda* 2, no. 9 (June 1, 1939): 61.

100. John W. Studebaker, "Scaling Cultural Frontiers," *Journal of Educational Sociology* 12, no. 8 (1939): 487–491; John W. Studebaker, *National Unity through Intercultural Education* (Washington, DC: US Government Printing Office, 1942).

101. Savage, *Broadcasting Freedom*, 36–37, 26.

102. Selig, *Americans All*, 243.

103. US Office of Education, *Americans All, Immigrants All* (Washington, DC, 1939), https://archive.org/details/americansallimmi00unit.

104. Savage, *Broadcasting Freedom*, 32.

105. Dan Shiffman, "A Standard for the Wise and Honest: The 'Americans All . . . Immigrants All' Radio Broadcasts," *Studies in Popular Culture* 19, no. 1 (1996): 99.

106. US Office of Education, *Americans All, Immigrants All*.

107. On the formation of the Radio Script Exchange, see Josh Shepperd, "Infrastructure in the Air: The Office of Education and the Development of Public Broadcasting in the United States, 1934–1944," *Critical Studies in Media Communication* 31, no. 3 (2014): 239–240. On mock broadcasts and other uses of the scripts in schools, see Hafemann, "Senn High School Intercultural Relations Laboratory"; Harold L. Ickes, *Annual Report of the Department of the Interior for the Fiscal Year Ended June 30, 1938* (Washington, DC: US Government Printing Office, 1938), 307–308; John W. Studebaker, *Annual Report of the United States Commissioner of Education for the Fiscal Year Ended June 30, 1940* (Washington, DC: US Government Printing Office, 1941), 87–88.

108. National Education Association, *Wartime Handbook for Education* (Washington, DC: National Education Association of the United States, 1943); Albert V. De Bonis, "Tolerance and Democracy: A Program for the English Class," *English Journal* 30, no. 2 (1941): 123–130; Hymen Alpern, "Brotherhood Week in the New York City High Schools," *School Review* 50, no. 6 (1942): 417–422; Hafemann, "Senn High School Intercultural Relations Laboratory," 40.

109. See Michele Hilmes, *Radio Voices: American Broadcasting, 1922–1952* (Minneapolis: University of Minnesota Press, 1997), 230–270.

110. The final installment of the film series, "War Comes to America," explored the idea of the United States as a nation of immigrants. Peter C. Rollins, "Frank Capra's Why We Fight Series and Our American Dream," *Journal of American Culture* 19, no. 4 (1996): 81–86.

111. Gerard Giordano, *Wartime Schools: How World War II Changed American Education* (New York: Peter Lang, 2004), 82.

112. Studebaker, *National Unity through Intercultural Education*; National Education Association, *Wartime Handbook for Education*, 18–19.

113. Burkholder, *Color in the Classroom*, 74–79. See also DuBois introduction in Ethel M. Duncan, *Democracy's Children* (New York: Hinds, Hayden and Eldgridge, Inc., 1945), xv.

114. Ruth Benedict and Gene Weltfish, *The Races of Mankind*, Public Affairs Pamphlet No. 85, Public Affairs Committee, Inc., 1943.

115. Hilmes, *Radio Voices*, 233, 251–252.

116. National Education Association, *Wartime Handbook for Education*, 18–19.

117. Rachel Davis DuBois, *National Unity through Intercultural Education* (Washington, DC: US Office of Education, 1942), 5.

118. DuBois, "Developing Sympathetic Attitudes toward Peoples," 388.

Chapter 5

1. The poem in the epigraph is quoted in Ellen McBryde Brown, "Interschool Correspondence Promotes International Understanding," *School Life* 13, no. 9 (1928): 172. Evaline Dowling, ed., *World Friendship: A Series of Articles Written by Some Teachers in the Los Angeles Schools and by a Few Others Who Are Likewise Interested in the Education of Youth* (Los Angeles: Committee on World Friendship, 1928); Suzanne Croizat Borghei, "Internationalism at the Grassroots: Los Angeles and Its City Schools, 1916–1953" (PhD diss., University of Southern California, 1995).

2. Lillie Newton Douglas, "An International Exchange Arrangement for Visual Aids," *Educational Screen* 7, no. 1 (1928): 11; Sidney L. Gulick, *Dolls of Friendship: The Story of a Goodwill Project between the Children of America and Japan* (New York: Committee on World Friendship among Children, 1929); Charles Roach, "Exhibits to and from Japan," in *World Friendship: A Series of Articles Written by Some Teachers in the Los Angeles Schools and by a Few Others Who Are Likewise Interested in the Education of Youth*, ed. Evaline Dowling (Los Angeles: Committee on World Friendship, 1928), 178–180.

3. Sheldon Himelfarb and Shamil Idriss, "Exchange 2.0" (Washington, DC: US Institute of Peace, 2011). The Stevens Initiative, founded in memory of US Ambassador to Libya J. Christopher Stevens in 2015, is a public-private partnership sponsored by the US Department of State and administered by the Aspen Institute, that awards grants to organizations that facilitate virtual exchange between youth in the US, the Middle East, and North Africa. Bureau of Educational and Cultural Affairs, "The Stevens Initiative," *Exchange Programs*, https://exchanges.state.gov/us/program/stevens-initiative.

4. Dowling, *World Friendship*, 174–180.

5. Katie Day Good, "Tracking Traveling Paper Dolls: New Media, Old Media, and Global Youth Engagement in the Flat Stanley Project," in *Civic Media: Technology, Design, Practice*, ed. Eric Gordon and Paul Mihailidis (Cambridge, MA: MIT Press, 2016), 421–428; Sarah Guth and Francesa Helm, eds., *Telecollaboration 2.0: Language,*

Literacies and Intercultural Learning in the 21st Century (Bern, Switzerland: Peter Lang, 2010); Robert O'Dowd, "Online Intercultural Exchanges," in *The Encyclopedia of Applied Linguistics*, ed. Carol A. Chapel (Hoboken, NJ: Wiley-Blackwell, 2012).

6. Richard Popp, "Machine-Age Communication: Media, Transportation, and Contact in Interwar America," *Technology and Culture* 52, no. 3 (2011): 459; Dowling, *World Friendship*, 102.

7. Popp, "Machine-Age Communication"; John Durham Peters, "Satan and Savior: Mass Communication in Progressive Thought," *Critical Studies in Mass Communication* 6 (1989): 247–263. For a study of the anxieties surrounding intercultural contact and emerging electric media in the late nineteenth century, see Carolyn Marvin, *When Old Technologies Were New: Thinking about Electric Communication in the Late Nineteenth Century* (New York: Oxford University Press, 1990), 191–231.

8. Florence Brewer Boeckel, *Through the Gateway* (New York: Macmillan, 1928), 92.

9. Herbert C. Hawk, "Winfield High's Homeroom Project in International Understanding," *Clearing House* 11, no. 3 (1936): 152.

10. Bettina Fabos and Michelle D. Young, "Telecommunication in the Classroom: Rhetoric versus Reality," *Review of Educational Research* 69, no. 3 (1999): 217–259.

11. Spencer Stoker, *The Schools and International Understanding* (Chapel Hill: University of North Carolina Press, 1933), 212–215.

12. Stoker, *Schools and International Understanding*, 214; Oscar Thiergen, "On International Correspondence between Pupils," *School Review* 7, no. 1 (1899): 4–10.

13. Edward H. Magill, "History of the International Correspondence in the United States of America," *Modern Language Notes* 17, no. 7 (1902): 228.

14. Magill, "History of the International Correspondence," 228.

15. Josephine C. Doniat, "International Correspondence of Pupils: Its History, Purpose, and Management," *School Review* 12, no. 1 (1904): 70–71, 77.

16. Thiergen, "International Correspondence between Pupils," 7.

17. Doniat, "International Correspondence of Pupils," 71; A. I. Roehm, "Organization and Educational Objectives of French-American Educational Correspondence," *French Review* 2, no. 1 (1928): 57.

18. Roehm, "Organization and Educational Objectives," 57; C. A. Wells, "Hobby Par Excellence!," *Rotarian* 46, no. 3 (1935): 2, 37.

19. Thomas Edward Oliver, "The National Peabody Foundation for International Educational Correspondence," *Modern Language Journal* 4, no. 2 (1919): 73–76.

20. Anonymous, "American Pupils Correspond with French Children," *School Life* 5, no. 7 (1920): 15.

21. Oliver, "National Peabody Foundation"; Elena Jackson Albarrán, *Seen and Heard in Mexico: Children and Revolutionary Cultural Nationalism* (Lincoln: University of Nebraska Press, 2015), 288–289.

22. Roehm, "Organization and Educational Objectives," 57.

23. P. P. Claxton, "American School Boys and Girls to Correspond with French," *School Life* 3, no. 6 (1919): 2, 8; Susan M. Dorsey, "Notices from Superintendent's Office," *Los Angeles School Journal* 3, no. 23 (1920): 21.

24. Charles M. Garnier, "A Call to the Schoolboys and Girls of America," *French Review* 2, no. 3 (1929): 247–249; Roehm, "Organization and Educational Objectives," 58–59.

25. Anonymous, "American Pupils Correspond with French Children."

26. Thomas Edward Oliver, "The National Peabody Foundation for International Educational Correspondence," *Modern Language Journal* 4, no. 2 (1919): 73–76.

27. Garnier, "Call to the Schoolboys and Girls of America," 248–249.

28. Stoker, *Schools and International Understanding*. Adding to the notion of correspondence as junior diplomacy, several international correspondences were arranged by American and foreign consuls. Dowling, *World Friendship*, 176.

29. Garnier, "Call to the Schoolboys and Girls of America," 249.

30. Roehm, "Organization and Educational Objectives," 57, 59.

31. A member of a Rotary Club chapter in Denmark, Sven Knudsen, organized letter and student exchanges among thousands of American and Scandinavian boys beginning in 1927. Sven Knudsen, "An International Holiday," *Rotarian* 31, no. 5 (November 1927): 30–31. For other international correspondence resources, see "Student Correspondence Information," *Phi Delta Kappan* 24, no. 3 (1941): 139; Bertha M. Bartholomew and C. L. Kulp, "Public-School Activities Designed to Develop Wholesome Nationalism and International Understanding," *Journal of Educational Sociology* 9, no. 7 (1936): 408–410.

32. Anonymous, "The Value of International Correspondence," *Boys' Life*, July 1933. See also the "World Brotherhood of Boys" column that appeared in the magazine in the 1920s and 1930s; Katharine Scherer Cronk, "Everyland Exchange," *Everyland* 7, no. 10 (1922): 31; Helen Mackintosh, "Pen-and-Ink Friendships for the Americas," *School Life*, July 1941, 297–298; Helen Mackintosh, "Building World Friendship through Correspondence," *Education for Victory* 3, no. 11 (December 4, 1944): 18, 23.

33. Hayumi Higuchi, "The Billiken Club: 'Race Leaders' Educating Children, 1921–1940," *Transforming Anthropology* 13, no. 2 (2005): 154–159. See, for example, Olive Sills, "Wants Pen Pals," *Chicago Defender*, May 28, 1932, 16; Billy Ennin, "African Will Form Branch Club for Us: Says He Is Fond of the Defender," *Chicago Defender*, October 10, 1936, 17.

34. "School News Digest," *Clearing House* 15, no. 3 (November 1, 1940): 178–192.

35. Edna MacDonough, "International Friendship by the Way of Youth," *World Affairs* 95, no. 3 (December 1, 1932): 169–170; Sara Fieldston, *Raising the World: Child Welfare in the American Century* (Cambridge, MA: Harvard University Press, 2015), 58, 93.

36. A. I. Roehm, "Learning Foreign Languages and Life by New Techniques Including International Educational Pupil-Correspondence," *Peabody Journal of Education* 19, no. 4 (1942): 227–228.

37. Mary MacDonald, "More Real Letters," *English Journal* 28, no. 9 (1939): 753. See also Hortense E. Braden, "Arsenal High Students' Letters Build World Pen-Friends," *Clearing House* 22, no. 7 (1948): 396–397; Adelaide L. Cunningham, "Corresponding with British Children," *English Journal* 34, no. 10 (1945): 560–562.

38. Mackintosh, "Pen-and-Ink Friendships"; Mackintosh, "Building World Friendship"; "Organizations for Cultural Exchange between the United States and Great Britain," *Education for Victory* 3, no. 11 (December 4, 1944): 23.

39. Joseph Burton Vasché, "10 Ideas for Timely Teaching of 1943 Social Studies: Stanislaus County Schools' Plan of Action," *Clearing House* 17 (1943): 471–473; Bartholomew and Kulp, "Public-School Activities," 409.

40. Liping Bu, "Educational Exchange and Cultural Diplomacy in the Cold War," *Journal of American Studies* 33, no. 3 (December 1, 1999): 393–415; Kevin V. Mulcahy, "Cultural Diplomacy and the Exchange Programs: 1938–1978," *Journal of Arts Management, Law and Society* 29, no. 1 (Spring 1999): 7; Christina Klein, *Cold War Orientalism: Asia in the Middlebrow Imagination, 1945–1961* (Berkeley: University of California Press, 2003); Fieldston, *Raising the World*.

41. World Goodwill Day was held on May 18, the anniversary of the First Hague Conference. Alongside chapter 4 in this book, see James F. Abel, "Goodwill Day—May 18," *School Life* 18, no. 9 (1933): 169.

42. Anonymous, "World Goodwill Day to Be Widely Observed," *Federal Council Bulletin*, 1927, 8; Linna Estelle Clark, *Promotion of International Goodwill through Education*, 1929, 14; E. O'Connor, "The World Brotherhood of Boys," *Boys' Life*, June 1927. Davies was a prominent Welsh pacifist who later helped found UNESCO. Wales for Peace, "Youth Message of Peace and Goodwill," 2015, http://www.walesforpeace.org/wfp/theme_peaceandgoodwill.html#TheMessage.

43. César Saerchinger, *Hello America! Radio Adventures in Europe* (Boston: Houghton Mifflin, 1938), 324–326. Thanks to Michael Socolow for bringing this source to my attention.

44. Lucille Fletcher, "Hands across the Sea," *Atlanta Constitution*, May 2, 1937.

45. On the tension between popular depictions of children as apolitical and political subjects, see Karen Dubinsky, "Children, Ideology, and Iconography: How Babies Rule the World," *Journal of the History of Childhood and Youth* 5, no. 1 (2012): 7–13.

46. Arthur H. Lynch and Willis K. Wing, "The International Radio Broadcast Test of 1924," *Radio Broadcasting* 6 (1924): 676–683; Bancroft Gherardi, "Voices across the Sea," *North American Review* 224, no. 838 (December 1, 1927): 654–661; James W. Carey and John J. Quirk, "The Mythos of the Electronic Revolution," *American Scholar* 39, no. 3 (1970): 395–424; Marvin, *When Old Technologies Were New*; Rachel Plotnick, "Touch of a Button: Long-Distance Transmission, Communication, and Control at World's Fairs," *Critical Studies in Media Communication* 30, no. 1 (2013): 52–68.

47. Anonymous, "Good-Will Chain to Mark Hague Day: International Telephone Conversations to Commemorate Founding," *Washington Post*, May 10, 1931; Anonymous, "Children of the World Will 'Meet' by Phone to Mark International Good-Will Day, May 15," *New York Times*, April 13, 1931, 27.

48. "Eleventh Annual Wireless Message of the Youth of Wales," 1932, Swarthmore Peace Collection, http://triptych.brynmawr.edu/cdm/ref/collection/SC_Ephemera/id/800.

49. Abel, "Goodwill Day—May 18."

50. Sarah Glassford, *Mobilizing Mercy: A History of the Canadian Red Cross* (Montreal: McGill-Queen's University Press, 2016), 93.

51. Woodrow Wilson, "Junior Red Cross," *Journal of Education* 86 (October 4, 1917): 319.

52. Henry Pomeroy Davison, *The American Red Cross in the Great War* (New York: Macmillan, 1919), 93–106; Anonymous, "Good Will to Men Reaches Farther Than Ever Before," *Red Cross Bulletin* 5, no. 2 (1921): 5; Philip Gibbs, "A World Wide League of Children," *Journal of Education* 93, no. 21 (1921): 563–565; Susan M. Strawn, *Knitting America: A Glorious Heritage from Warm Socks to High Art* (London: Voyageur Press, 2011).

53. Douglas Griesemer, "Shop Talk: International Correspondence of the Junior Red Cross," *Elementary English Review* 16, no. 7 (1939): 284–285.

54. Arthur William Dunn, "Exchange of Letters Aids Language Study," *Journal of Education* 103, no. 21 (1926): 574–575; Brown, "Interschool Correspondence";

Everett Baxter Sackett, "The Administration of the International School Correspondence of the Junior Red Cross" (PhD diss., Teachers College, Columbia University, 1931), 6; Stoker, *Schools and International Understanding*, 212; Alma Barker and Helen L. Chambers, "International Correspondence via the Junior Red Cross," *English Journal* 40, no. 5 (1951): 279–280.

55. Dunn, "Exchange of Letters Aids Language Study."

56. Stoker, *Schools and International Understanding*, 218; Sackett, "Administration of the International School Correspondence," 16–17.

57. Annmarie Valdes, "'I, Being a Member of the Junior Red Cross, Gladly Offered My Services': Transnational Practices of Citizenship by the International Junior Red Cross Youth," *Transnational Social Review* 5, no. 2 (2015): 161–175.

58. Dunn, "Exchange of Letters Aids Language Study," 574.

59. Gibbs, "World Wide League of Children," 563–565; Martin C. Kerby, *Sir Philip Gibbs and English Journalism in War and Peace* (London: Palgrave Macmillan, 2016), 156.

60. Stoker, *Schools and International Understanding*.

61. One JRC official described the albums as "illustrated letters, bound in volumes, and other materials [that] cover topics that interest entire groups." Griesemer, "Shop Talk," 284. See also Dowling, *World Friendship*, 127.

62. Sample titles prepared by a high school in Massachusetts included *Holiday Time in America*, *Sports in the USA*, and *American Summer*. Louise B. Forsyth, "We Correspond with the World," *English Journal* 46, no. 9 (1957): 556–558.

63. For one Washington, DC, teacher, this collaborative form of international correspondence was ideal for enlarging the educational value of international exchanges beyond traditional international correspondence, as it "influences more boys and girls than correspondence between individuals." Barker and Chambers, "International Correspondence," 278.

64. Zilla Wiswall, "The Junior Red Cross as a Motivating Force in English: International School Correspondence," *Elementary English Review* 6, no. 6 (1929): 155.

65. Dunn, "Exchange of Letters Aids Language Study," 474–475.

66. Hawk, "Winfield High's Homeroom Project," 152.

67. Brown, "Interschool Correspondence," 172.

68. Henry Lester Smith and Peyton Henry Canary, "Some Practical Efforts to Teach Good Will," *Bulletin of the Indiana University School of Education* 6, no. 4 (1935): 78–80.

69. Sackett, "Administration of the International School Correspondence," 30–35. Surviving scrapbooks and administrative forms preserved in archives attest to the handmade nature of JRC correspondence. Schools were instructed to use covers that were "made, not bought," and their contents were each entirely unique, filled with student-made essays, sketches, paintings, samples of handiwork, collected natural and industrial specimens, and everyday objects like buttons, coins, and badges. See "Junior Red Cross Scrapbooks," n.d., American National Red Cross Mile High Chapter Collection, boxes 4 and 7, OV folios 12–16, Denver Public Library; "Junior Red Cross Scrapbooks, 1945–1954," RG 14 Laboratory School Records, S3 Scrapbooks and Awards/Clippings, University of Northern Colorado Archives.

70. Sackett, "Administration of the International School Correspondence," 51–57.

71. Calanthe M. Brazelton, "Tucson High Goes to Town on the Junior Red Cross," *Clearing House* 17, no. 3 (1942): 157.

72. Brazelton, "Tucson High Goes to Town," 157.

73. Julia F. Irwin, *Making the World Safe: The American Red Cross and a Nation's Humanitarian Awakening* (New York: Oxford University Press, 2013); Fieldston, *Raising the World*.

74. "Children's Crusade for Children," 1940, Children's Crusade for Children Records, Swarthmore Peace Collection.

75. Schools of Penn Yan, New York, "Children's Crusade," April 1940, Children's Crusade for Children Records, 1940s, box 23, folder: Publicity, Library of Congress, Manuscript Division.

76. "Hoboken Junior Red Cross Scrapbook," 1940, Children's Crusade for Children Records, OV box 2, Library of Congress, Manuscript Division.

77. "Children's Crusade for Children: Special Edition of SOWCHS," Westport Central High School, 1940, Children's Crusade for Children Records, Swarthmore Peace Collection.

78. Helena High School, "Hands across the Sea." *Nugget*, Helena, Montana, April 19, 1940, Children's Crusade for Children Records, Swarthmore Peace Collection.

79. "True, there are many serious problems in America which warrant our attention, but it must be remembered that in the present crisis of world affairs America remains not only the most democratic but the wealthiest of all the nations." Harold Schiffrin, "Address to the School by Student Chairman of the Crusade," Rochester Public Schools, April 1940, Children's Crusade for Children Records, 1940s, box 23, folder: Publicity, Library of Congress, Manuscript Division. See also "'Brothers': Playlet Written by 7A Grade of #30 School for Radio Program," Rochester Public Schools, April 1940, Children's Crusade for Children Records, 1940s, box 23, folder: Publicity, Library of Congress, Manuscript Division.

80. Committee on World Friendship among Children, "CWFC Minutes, November 22, 1926," 1926, Federal Council of the Churches of Christ in America Records, 1894–1952, box 44, folder 21, Presbyterian Historical Society.

81. Committee on World Friendship among Children, "The Friendship Project for China," 1932, Federal Council of the Churches of Christ in America Records, 1894–1952, box 44, folder 22, Presbyterian Historical Society; Committee on World Friendship among Children, "CWFC Minutes, December 19, 1934," 1934, 19, Federal Council of the Churches of Christ in America Records, 1894–1952, box 44, folder 22, Presbyterian Historical Society.

82. Jeannette Emrich, "Packing Friendship into School Bags," *Federal Council Bulletin*, 1928, Federal Council of the Churches of Christ in America Records, 1894–1952, box 44, folder 21, Presbyterian Historical Society.

83. Committee on World Friendship among Children, "CWFC Minutes, June 15, 1926," 1926, 15, Federal Council of the Churches of Christ in America Records, 1894–1952, box 44, folder 21, Presbyterian Historical Society.

84. Committee on World Friendship among Children, "Doll Messengers of Friendship," 1926, Federal Council of the Churches of Christ in America Records, 1894–1952, box 44, folder 21, Presbyterian Historical Society.

85. Emrich, "Packing Friendship into School Bags."

86. Jeannette Emrich, "A Half-Million Smiles: Reports from Philippines Record Thrilling Reception of Treasure Chest," *Federal Council Bulletin*, 1931, Federal Council of the Churches of Christ in America Records, 1894–1952, box 44, folder 21, Presbyterian Historical Society.

87. Wilbur Zelinsky, "The Twinning of the World: Sister Cities in Geographic and Historical Perspective," *Annals of the Association of American Geographers* 81, no. 1 (1991): 1–31.

88. Committee on International Friendship among Children in Japan, *Welcome to the American Doll-Messengers* (Tokyo: Herald of Asia, 1927), 2; Gulick, *Dolls of Friendship*, 63, 69.

89. Anonymous, "Mexico Sends Appreciation of Good-Will Gifts," *New York Herald Tribune*, April 13, 1929.

90. Irwin, *Making the World Safe*.

91. Anne McClintock, *Imperial Leather: Race, Gender, and Sexuality in the Colonial Contest* (New York: Routledge, 1995).

92. Committee on World Friendship among Children, "Summary of Correspondence regarding the next Project," 1931, 1, Federal Council of the Churches of Christ in America Records, 1894–1952, box 44, folder 21, Presbyterian Historical Society.

93. Committee on World Friendship among Children, "Doll Messengers of Friendship," Federal Council of the Churches of Christ in America Records, 1894–1952, box 44, folder 21, Presbyterian Historical Society. On the race of dolls, see CWFC minutes, June 15, 1926, Federal Council of the Churches of Christ in America Records, 1894–1952, box 44, folder 21, Presbyterian Historical Society.

94. Albarrán writes in greater detail on the geopolitical undertones of the CWFC shipment of goodwill bags to Mexico, arguing that they "reinforced racialized Good Neighbor-era policies that established the United States as the cultural imperialist power, a father figure extending health and well-being down to an infantilized nascent democracy," along with the stereotype of "Mexico as infantile, Indian land requiring U.S. charity to transform its residents into productive citizens." Albarrán, *Seen and Heard in Mexico*, 248–249.

95. Anonymous, "Mexican Arts and Crafts on Exhibition at Y.W.C.A.," *Atlanta Constitution*, April 6, 1930.

96. Committee on International Friendship among Children in Japan, *Welcome to the American Doll-Messengers*, 2–6.

97. Fieldston, *Raising the World*, 4.

Chapter 6

1. Marshall McLuhan and National Association of Educational Broadcasters, *Report on Project in Understanding New Media* (Urbana, IL: US Office of Education, June 30, 1960); Edgar Dale, "What Is the Image of the Man of Tomorrow?," 1960, Edgar Dale Papers, box 7, folder 158, Ohio State University; William J. Clinton, "The President's Radio Address," American Presidency Project, February 8, 1997, http://www.presidency.ucsb.edu/ws/index.php?pid=53524.

2. Deneen Frazier, *Internet for Kids* (Alameda, CA: SYBEX, Inc., 1996), 29.

3. William J. Clinton, "Statement on Signing the Telecommunications Act of 1996," American Presidency Project, February 8, 1996, accessed July 24, 2018, http://www.presidency.ucsb.edu/ws/?pid=52289.

4. Barack Obama, "Remarks in Cairo," American Presidency Project, June 4, 2009, http://www.presidency.ucsb.edu/ws/index.php?pid=86221; US Department of Education, "International Engagement through Education: Remarks by Secretary Arne Duncan at the Council on Foreign Relations Meeting," May 26, 2010, http://www.ed.gov/news/speeches/international-engagement-through-education-remarks-secretary-arne-duncan-council-forei.

5. Marshall McLuhan, *Understanding Media: The Extensions of Man* (Cambridge, MA: MIT Press, 1994), 3, 5. While McLuhan elaborated on the concept of the globe as a

village in *Understanding Media*, the term "global village" appeared in his earlier work. "The new electronic interdependence recreates the world in the image of a global village." Marshall McLuhan, *The Gutenberg Galaxy: The Making of Typographic Man* (Toronto: University of Toronto Press, 1962), 31. See also McLuhan and National Association of Educational Broadcasters, *Report on Project in Understanding New Media*, 2.

6. John M. Carroll and Mary Beth Rosson, "The Neighborhood School in the Global Village," *IEEE Technology and Society Magazine* 17, no. 4 (1998): 4. See also Ernest L. Boyer, *Cornerstones for a New Century: Teacher Preparation, Early Childhood Education, a National Education Index* (Washington, DC: NEA Professional Library, National Education Association, 1992); David D. Thornburg, "Technology in K–12 Education: Envisioning a New Future," 1999, http://eric.ed.gov/?id=ED452843; George Kontos and Al P. Mizell, "Global Village Classroom: The Changing Roles of Teachers and Students through Technology," *TechTrends* 42, no. 5 (1997): 17–22.

7. Charles R. Acland, "Never Too Cool for School," *Journal of Visual Culture* 13, no. 1 (2014): 13; Fred Turner, *The Democratic Surround: Multimedia and American Liberalism from World War II to the Psychedelic Sixties* (Chicago: University of Chicago Press, 2013), 1, 272–274.

8. McLuhan and National Association of Educational Broadcasters, *Report on Project in Understanding New Media*; Acland, "Never Too Cool for School"; James W. Carey, "Marshall McLuhan: Genealogy and Legacy," *Canadian Journal of Communication* 23, no. 3 (1998), http://www.cjc-online.ca/index.php/journal/article/view/1045.

9. Marshall McLuhan and George B. Leonard, "The Future of Education: The Class of 1989," *LOOK*, February 21, 1967; McLuhan and National Association of Educational Broadcasters, *Report on Project in Understanding New Media*, 9.

10. Paul Saettler, *The Evolution of American Educational Technology* (Greenwich, CT: Information Age Publishing, 2004), 167–168. On membership numbers, see "DVI/ DAVI/AECT: Number of Members, by Decade," Association for Educational Communications and Technology, 2014, http://c.ymcdn.com/sites/aect.site-ym.com/ resource/resmgr/about_pdf/memchart.pdf.

11. J. G. Umstattd, "Education for Peace," *High School Journal* 30, no. 1 (1947): 19–27.

12. McLuhan and National Association of Educational Broadcasters, *Report on Project in Understanding New Media*, 3; Charles R. Acland, "Curtains, Carts and the Mobile Screen," *Screen* 50, no. 1 (2009): 161.

13. Saettler, *Evolution of American Educational Technology*, 372–378.

14. Stephen J. Feit, "Vive la technologie!," *Educational Screen* 41, no. 11 (1962): 646–649.

15. Encyclopaedia Britannica, "Leadership in Audio-Visual Materials for Space Age Education," *Educational Screen* 38, no. 1 (1959): 28–29.

16. Paul S. Welty, "The Role of Audiovisual Associations in International Relations," *Audio Visual Communication Review* 9, no. 6 (1961): 306–312; Ruth B. Hedges, "A Visual Education Program for UNESCO," *Hollywood Quarterly* 2, no. 1 (1946): 97–99.

17. World Confederation of Organizations of the Teaching Profession, *Audio-Visual Aids for International Understanding* (Washington, D.C.: World Confederation of Organizations of the Teaching Profession, 1961); Anonymous, "Keeping Abreast in Education," *Phi Delta Kappan* 42, no. 9 (1961): 413–416.

18. F. Dean McClusky, *Audio-Visual Teaching Techniques* (Dubuque, IA: Wm. C. Brown Company, 1950); Richard H. Heindel, "The American Library Abroad: A Medium of International Intellectual Exchange," *Library Quarterly* 16, no. 2 (1946): 93–107.

19. Victoria Cain, "From *Sesame Street* to Prime Time School Television: Educational Media in the Wake of the Coleman Report," *History of Education Quarterly* 57, no. 4 (November 2017): 590–601; Wilbur Schramm, "Instructional Television: Promise and Opportunity" (National Association of Educational Broadcasters, January 1967); Dean T. Jamison, Patrick Suppes, and Christopher Butler, "Estimated Costs of Computer Assisted Instruction for Compensatory Education in Urban Areas," *Educational Technology* 10, no. 9 (1970): 49–57; Larry Cuban, *Teachers and Machines: The Use of Classroom Technology since 1920* (New York: Teachers College Press, 1986); Brian Goldfarb, *Visual Pedagogy: Media Cultures in and beyond the Classroom* (Durham, NC: Duke University Press, 2002); Saettler, *Evolution of American Educational Technology*.

20. Saettler, *Evolution of American Educational Technology*; Cuban, *Teachers and Machines*, 31.

21. Cuban, *Teachers and Machines*, 30.

22. Saettler, *Evolution of American Educational Technology*, 370; Cuban, *Teachers and Machines*, 32; Wilbur Lang Schramm, Lyle M. Nelson, and Mere T. Betham, *Bold Experiment: The Story of Educational Television in American Samoa* (Palo Alto, CA: Stanford University Press, 1981).

23. Audrey Watters, "Teaching by Television in American Samoa: A History," *Hack Education* (blog), June 6, 2015, http://hackeducation.com/2015/06/06/american-samoa-educational-tv.

24. Edgar Dale, "A Proposal for International Audiovisual Exchange," *Audio Visual Communication Review* 8, no. 4 (1960): 230; McClusky, *Audio-Visual Teaching Techniques*, 105–108.

25. Goldfarb, *Visual Pedagogy*, 19. The term *mediascapes* was coined by Arjun Appadurai to "refer both to the distribution of the electronic capabilities to produce and

disseminate information (newspapers, magazines, television stations, and film-production studios), which are now available to a growing number of private and public interests throughout the world, and to the images of the world created by these media." Arjun Appadurai, *Modernity at Large: Cultural Dimensions of Globalization* (Minneapolis: University of Minnesota Press, 1996), 35.

26. Tessa Melvin, "A Computer Project Links Students with Soviet Partners," *New York Times*, December 4, 1988; Barton Reppert, "Computer Age Glasnost: Satellite Links up Americans and Soviets via Personal Computers," *Los Angeles Times*, January 14, 1990; Boris Berenfeld, "Linking East-West Schools via Telecomputing," *THE Journal (Technological Horizons in Education)* 20, no. 6 (1993); Bruce Watson, "The Wired Classroom: American Education Goes On-Line," *Phi Delta Kappan* 72, no. 2 (October 1, 1990): 109–112; Barbara Hof, "From Harvard via Moscow to West Berlin: Educational Technology, Programmed Instruction and the Commercialisation of Learning after 1957," *History of Education* 47, no. 4 (2018): 445–465.

27. Melvin, "Computer Project Links Students with Soviet Partners"; Watson, "Wired Classroom"; Reppert, "Computer Age Glasnost"; Berenfeld, "Linking East-West Schools via Telecomputing"; "IEARN | Learning with the World, Not Just about It," iEARN, http://www.iearn.org.

28. Dale Hubert, "The Flat Stanley Project and Other Authentic Applications of Technology in the Classroom," in *Innovative Approaches to Literacy Education: Using the Internet to Support New Literacies*, ed. Rachel A. Karchmer, Marla H. Mallette, Julia Kara-Soteriou, and Donald J. Leu (Newark, DE: International Reading Association, 2005); Katie Day Good, "Tracking Traveling Paper Dolls: New Media, Old Media, and Global Youth Engagement in the Flat Stanley Project," in *Civic Media: Technology, Design, Practice*, ed. Eric Gordon and Paul Mihailidis (Cambridge, MA: MIT Press, 2016), 421–428.

29. Michalinos Zembylas and Charalambos Vrasidas, "Globalization, Information and Communication Technologies, and the Prospect of a 'Global Village': Promises of Inclusion or Electronic Colonization?," *Journal of Curriculum Studies* 37, no. 1 (2005): 65–83.

30. Kathryn Toure, Mamadou Lamine Diarra, Thierry Karsenti, and Salomon Tchaméni-Ngamo, "Reflections on Cultural Imperialism and Pedagogical Possibilities Emerging from Youth Encounters with the Internet in Africa," in *ICT and Changing Mindsets in Education*, ed. Kathryn Toure, Therese Mungah Shalo Tchombe, and Thierry Karsenti (Bamenda, Cameroon: Langaa Research and Publishing Common Initiative Group, 2008), 7; Morgan G. Ames, "Learning Consumption: Media, Literacy, and the Legacy of One Laptop per Child," *Information Society* 32, no. 2 (March 2016): 85–97.

31. Bettina Fabos and Michelle D. Young, "Telecommunication in the Classroom: Rhetoric versus Reality," *Review of Educational Research* 69, no. 3 (1999): 217–259.

32. Anna Lauren Hoffmann, Nicholas Proferes, and Michael Zimmer, "'Making the World More Open and Connected': Mark Zuckerberg and the Discursive Construction of Facebook and Its Users," *New Media and Society* 20, no. 1 (January 2018): 199–218; Josh Constine, "Facebook Changes Mission Statement to 'Bring the World Closer Together,'" *TechCrunch* (blog), http://social.techcrunch.com/2017/06/22/bring -the-world-closer-together/.

33. Natasha Singer and Danielle Ivory, "How Silicon Valley Plans to Conquer the Classroom," *New York Times*, November 3, 2017, Technology section, https://www .nytimes.com/2017/11/03/technology/silicon-valley-baltimore-schools.html.

34. "Bring Your Lessons to Life with Expeditions," Google, 2018, https://www .google.com/expeditions/; "MysterySkype," Skype in the Classroom, Microsoft, https://education.microsoft.com/skype-in-the-classroom/mystery-skype.

35. Lev Grossman, "Inside Facebook's Plan to Wire the World," *Time*, December 15, 2014, http://time.com/facebook-world-plan/.

36. Ramesh Srinivasan, *Whose Global Village?: Rethinking How Technology Shapes Our World* (New York: NYU Press, 2017); Fabos and Young, "Telecommunication in the Classroom."

37. Nigel Thrift, "New Urban Eras and Old Technological Fears: Reconfiguring the Goodwill of Electronic Things," *Urban Studies* 33, no. 8 (1996): 1463–1493; Neil Selwyn, "Looking beyond Learning: Notes towards the Critical Study of Educational Technology," *Journal of Computer Assisted Learning* 26, no. 1 (2010): 65–73.

38. Jennifer S. Light, "Rethinking the Digital Divide," *Harvard Educational Review* 71, no. 4 (2001): 709–734; Christo Sims, *Disruptive Fixation: School Reform and the Pitfalls of Techno-Idealism* (Princeton, NJ: Princeton University Press, 2017).

39. Bettina Fabos, *Wrong Turn on the Information Superhighway: Education and the Commercialization of the Internet* (New York: Teachers College Press, 2004), 4.

40. Larry Cuban, *Oversold and Underused: Computers in the Classroom* (Cambridge, MA: Harvard University Press, 2003); Todd Oppenheimer, *The Flickering Mind: Saving Education from the False Promise of Technology* (New York: Random House, 2007).

41. Charles Ess, "Computer-Mediated Colonization, the Renaissance, and Educational Imperatives for an Intercultural Global Village," *Ethics and Information Technology* 4, no. 1 (2002): 11–22; Paul C. Gorski, "Multicultural Education and Progressive Pedagogy in the Online Information Age," *Multicultural Perspectives* 6, no. 4 (2004): 37–48; Zembylas and Vrasidas, "Globalization, Information and Communication Technologies"; Gail E. Hawisher and Cynthia L. Selfe, eds., *Global Literacies and the World Wide Web* (New York: Routledge, 2005); Nicole Starosielski, *The Undersea Network* (Durham, NC: Duke University Press, 2015).

42. Safiya Umoja Noble, *Algorithms of Oppression: How Search Engines Reinforce Racism* (New York: NYU Press, 2018).

43. Ess, "Computer-Mediated Colonization, the Renaissance, and Educational Imperatives for an Intercultural Global Village," 12.

44. Zembylas and Vrasidas, "Globalization, Information and Communication Technologies," 66.

45. Mark Warschauer, *Technology and Social Inclusion: Rethinking the Digital Divide* (Cambridge, MA: MIT Press, 2004), 6. See also Henry Jenkins, Katie Clinton, Ravi Purushotma, Alice J. Robison, and Margaret Weigel, *Confronting the Challenges of Participatory Culture: Media Education for the 21st Century*, (Cambridge, MA: MIT Press, 2009).

46. Roderic N. Crooks, "Times Thirty: Access, Maintenance, and Justice," *Science, Technology, and Human Values* (2018). See also Ames, "Learning Consumption"; Virginia Eubanks, *Digital Dead End: Fighting for Social Justice in the Information Age* (Cambridge, MA: MIT Press, 2012).

47. The literature on this is extensive, but see, for example, Karen E. Wohlwend and Cynthia Lewis, "Critical Literacy, Critical Engagement, and Digital Technology: Convergence and Embodiment in Global Spheres," in *Handbook of Research on Teaching the English Language Arts*, ed. Diane Lapp and Douglas Fisher (New York: Routledge, 2011), 188–194; Mizuko Ito, Kris Gutiérrez, Sonia Livingstone, Bill Penuel, Jean Rhodes, Katie Salen, Juliet Schor, Julian Sefton-Green, and S. Craig Watkins, *Connected Learning: An Agenda for Research and Design* (Irvine, CA: Digital Media and Learning Research Hub, 2013); Sonia Livingstone and Julian Sefton-Green, *The Class: Living and Learning in the Digital Age* (New York: NYU Press, 2016).

48. Whitney Phillips, *This Is Why We Can't Have Nice Things: Mapping the Relationship between Online Trolling and Mainstream Culture* (Cambridge, MA: MIT Press, 2015); Alice Marwick and Rebecca Lewis, *Media Manipulation and Disinformation Online* (New York: Data and Society Research Institute, 2017).

49. Jeet Heer, "We're Going to See a Lot More Walls," *New Republic*, April 24, 2018, https://newrepublic.com/article/148109/were-going-see-lot-walls.

50. Stella S. Center, "The Responsibility of Teachers of English in Contemporary American Life," *English Journal* 22, no. 2 (1933): 97–108.

51. Joseph Kahne and Ellen Middaugh, "Democracy for Some: The Civic Opportunity Gap in High School" (working paper 59, Center for Information and Research on Civic Learning and Engagement, February 2008); Michele S. Moses and John Rogers, "Enhancing a Nation's Democracy through Equitable Schools," in *Closing the Opportunity Gap: What America Must Do to Give Every Child an Even Chance*, ed.

Prudence L. Carter and Kevin G. Welner (New York: Oxford University Press, 2013), 207–216.

52. Cheryl E. Matias, *Feeling White: Whiteness, Emotionality, and Education* (Rotterdam: Sense Publishers, 2016); Emily Deruy, "Student Diversity Is Up but Teachers Are Mostly White," American Association of Colleges for Teacher Education, https://aacte.org/news-room/aacte-in-the-news/347-student-diversity-is-up-but-teachers-are-mostly-white.

53. James A. Banks, "Diversity, Group Identity, and Citizenship Education in a Global Age," *Educational Researcher* 37, no. 3 (2008): 129–139.

54. Talya Zemach-Bersin, "Global Citizenship and Study Abroad: It's All about U.S.," *Critical Literacy: Theories and Practices* 1 (2007): 20.

55. Vanessa de Oliveira Andreotti and Lynn Mario T. M. de Souza, eds., *Postcolonial Perspectives on Global Citizenship Education* (New York: Routledge, 2014); Banks, "Diversity, Group Identity, and Citizenship Education in a Global Age."

56. Katharyne Mitchell, "Educating the National Citizen in Neoliberal Times: From the Multicultural Self to the Strategic Cosmopolitan," *Transactions of the Institute of British Geographers* 28, no. 4 (2003): 387–403; Julie Matthews and Ravinder Sidhu, "Desperately Seeking the Global Subject: International Education, Citizenship and Cosmopolitanism," *Globalisation, Societies, and Education* 3 (2005): 49–66; Marzia Cozzolino DiCicco, "Global Citizenship Education within a Context of Accountability and 21st Century Skills: The Case of Olympus High School," *Education Policy Analysis Archives* 24 (2016): 57; Catherine Hartung, "Global Citizenship Incorporated: Competing Responsibilities in the Education of Global Citizens," *Discourse: Studies in the Cultural Politics of Education* 38, no. 1 (January 2, 2017): 16–29.

57. Henry Jenkins, Sangita Shresthova, Liana Gamber-Thompson, Neta Kugler-Vilenchik, and Arely Zimmerman, *By Any Media Necessary: The New Youth Activism* (New York: NYU Press, 2016); Jean E. Burgess, Marcus Foth, and Helen G. Klaebe, "Everyday Creativity as Civic Engagement: A Cultural Citizenship View of New Media," in *Communications Policy and Research Forum* (Sydney, 2006), http://eprints.qut.edu.au/5056/.

58. John Dewey, "The School as Social Center," *Elementary School Teacher* 3, no. 2 (1902): 73–86; John Dewey, *Democracy and Education: An Introduction to the Philosophy of Education* (New York: Macmillan, 1916).

59. Shirin Vossoughi, Paula K. Hooper, and Meg Escudé, "Making through the Lens of Culture and Power: Toward Transformative Visions for Educational Equity," *Harvard Educational Review* 86, no. 2 (2016): 206–232; Matt Ratto and Megan Boler, eds., *DIY Citizenship: Critical Making and Social Media* (Cambridge, MA: MIT Press, 2014).

Index